TSI Subject Test

Mathematics

Student Practice Workbook

+ Two Full-Length TSI Tests

Math Notion

www.MathNotion.com

TSI Subject Test Mathematics

Published in the United State of America By

The Math Notion

Web: WWW.MathNotion.com

Email: info@Mathnotion.com

ISBN: 978-1-63620-051-4

The Math Notion

Michael Smith has been a math instructor for over a decade now. He launched the Math Notion. Since 2006, we have devoted our time to both teaching and developing exceptional math learning materials. As a test prep company, we have worked with thousands of students. We have used the feedback of our students to develop a unique study program that can be used by students to drastically improve their math scores fast and effectively. We have more than a thousand Math learning books including:

– **SAT Math Prep**

– **ACT Math Prep**

– **Accuplacer Math Prep**

– **College Mathematics workbook**

– **CLEP College Algebra workbook**

–**many Math Education Workbooks, Study Guides, Practice and Exercise Books**

As an experienced Math test preparation company, we have helped many students raise their standardized test scores—and attend the colleges of their dreams: We tutor online and in person, we teach students in large groups, and we provide training materials and textbooks through our website and through Amazon.

You can contact us via email at:

info@Mathnotion.com

Get the Targeted Practice You Need to Ace the TSI Math Test!

TSI Subject Test - Mathematics includes easy-to-follow instructions, helpful examples, and plenty of math practice problems to assist students to master each concept, brush up their problem-solving skills, and create confidence.

The TSI math practice book provides numerous opportunities to evaluate basic skills along with abundant remediation and intervention activities. It is a skill that permits you to quickly master intricate information and produce better leads in less time. Students can boost their test-taking skills by taking the book's two practice TSI Math exams. All test questions answered and explained in detail.

Important Features of the TSI Math Book:

- A **complete review** of TSI math test topics,
- Over 2,500 practice problems covering all topics tested,
- The most important concepts you need to know,
- Clear and concise, easy-to-follow sections,
- Well designed for enhanced learning and interest,
- Hands-on experience with all question types
- **2 full-length practice tests** with detailed answer explanations
- Cost-Effective Pricing

Powerful math exercises to help you avoid traps and pacing yourself to beat the TSI test. Students will gain valuable experience and raise their confidence by taking math practice tests, learning about test structure, and gaining a deeper understanding of what is tested on the TSI Math. If ever there was a book to respond to the pressure to increase students' test scores, this is it.

WWW.MathNotion.COM

… So Much More Online!

✓ FREE Math Lessons

✓ More Math Learning Books!

✓ Mathematics Worksheets

✓ Online Math Tutors

For a PDF Version of This Book

Please Visit WWW.MathNotion.com

Contents

Chapter 1 :

Integers and Number Theory

Topics that you will practice in this chapter:

- ✓ Rounding
- ✓ Rounding and Estimates
- ✓ Adding and Subtracting Integers
- ✓ Multiplying and Dividing Integers
- ✓ Order of Operations
- ✓ Ordering Integers and Numbers
- ✓ Integers and Absolute Value
- ✓ Factoring Numbers
- ✓ Greatest Common Factor (GCF)
- ✓ Least Common Multiple (LCM)
- ✓ Sets

"Wherever there is number, there is beauty." −Proclus

Rounding

✎ Round each number to the nearest ten.

1) 42 = ____   5) 19 = ____   9) 48 = ____

2) 88 = ____   6) 25 = ____   10) 81 = ____

3) 24 = ____   7) 93 = ____   11) 58 = ____

4) 57 = ____   8) 71 = ____   12) 87 = ____

✎ Round each number to the nearest hundred.

13) 198 = ____   17) 321 = ____   21) 580 = ____

14) 387 = ____   18) 433 = ____   22) 868 = ____

15) 816 = ____   19) 579 = ____   23) 480 = ____

16) 101 = ____   20) 825 = ____   24) 287 = ____

✎ Round each number to the nearest thousand.

25) 1,382 = ____   29) 9,099 = ____   33) 52,866 = ____

26) 3,420 = ____   30) 22,980 = ____   34) 85,190 = ____

27) 4,254 = ____   31) 45,188 = ____   35) 70,990 = ____

28) 6,861 = ____   32) 16,808 = ____   36) 26,869 = ____

Rounding and Estimates

✎ **Estimate the sum by rounding each number to the nearest ten.**

1) $13 + 22 =$ _____

2) $71 + 23 =$ _____

3) $61 + 58 =$ _____

4) $56 + 85 =$ _____

5) $368 + 249 =$ _____

6) $330 + 903 =$ _____

7) $471 + 293 =$ _____

8) $1,950 + 2,655 =$ _____

✎ **Estimate the product by rounding each number to the nearest ten.**

9) $32 \times 71 =$ _____

10) $12 \times 33 =$ _____

11) $31 \times 83 =$ _____

12) $19 \times 11 =$ _____

13) $42 \times 76 =$ _____

14) $63 \times 34 =$ _____

15) $19 \times 31 =$ _____

16) $59 \times 71 =$ _____

✎ **Estimate the sum or product by rounding each number to the nearest ten.**

17) $\begin{array}{r} 29 \\ \times\ 12 \\ \hline \end{array}$

18) $\begin{array}{r} 37 \\ \times\ 26 \\ \hline \end{array}$

19) $\begin{array}{r} 48 \\ +\ 82 \\ \hline \end{array}$

20) $\begin{array}{r} 65 \\ +44 \\ \hline \end{array}$

21) $\begin{array}{r} 37 \\ \times\ 14 \\ \hline \end{array}$

22) $\begin{array}{r} 71 \\ +\ 32 \\ \hline \end{array}$

Adding and Subtracting Integers

✏ **Find each sum.**

1) $14 + (-6) =$

2) $(-13) + (-20) =$

3) $5 + (-28) =$

4) $50 + (-12) =$

5) $(-7) + (-15) + 3 =$

6) $30 + (-14) + 8 =$

7) $40 + (-10) + (-14) + 17 =$

8) $(-15) + (-20) + 13 + 35 =$

9) $40 + (-20) + (38 - 29) =$

10) $28 + (-12) + (30 - 12) =$

✏ **Find each difference.**

11) $(-18) - (-7) =$

12) $25 - (-14) =$

13) $(-20) - 36 =$

14) $34 - (-19) =$

15) $51 - (30 - 21) =$

16) $17 - (5) - (-24) =$

17) $(35 + 20) - (-46) =$

18) $48 - 16 - (-8) =$

19) $62 - (28 + 17) - (-15) =$

20) $58 - (-23) - (-31) =$

21) $19 - (-8) - (-13) =$

22) $(19 - 24) - (-14) =$

23) $27 - 33 - (-21) =$

24) $58 - (32 + 24) - (-9) =$

25) $36 - (-30) + (-17) =$

26) $27 - (-42) + (-31) =$

Multiplying and Dividing Integers

✎ **Find each product.**

1) $(-9) \times (-5) =$

2) $(-3) \times 9 =$

3) $8 \times (-12) =$

4) $(-7) \times (-20) =$

5) $(-3) \times (-5) \times 6 =$

6) $(14 - 3) \times (-8) =$

7) $12 \times (-9) \times (-3) =$

8) $(140 + 10) \times (-2) =$

9) $10 \times (-12 + 8) \times 3 =$

10) $(-8) \times (-5) \times (-10) =$

✎ **Find each quotient.**

11) $42 \div (-7) =$

12) $(-48) \div (-6) =$

13) $(-40) \div (-8) =$

14) $54 \div (-2) =$

15) $152 \div 19 =$

16) $(-144) \div (-12) =$

17) $180 \div (-10) =$

18) $(-312) \div (-12) =$

19) $221 \div (-13) =$

20) $(-126) \div (6) =$

21) $(-161) \div (-7) =$

22) $-266 \div (-14) =$

23) $(-120) \div (-4) =$

24) $270 \div (-18) =$

25) $(-208) \div (-8) =$

26) $(135) \div (-15) =$

Order of Operations

✎ **Evaluate each expression.**

1) $7 + (5 \times 4) =$

2) $14 - (3 \times 6) =$

3) $(19 \times 4) + 16 =$

4) $(16 - 7) - (8 \times 2) =$

5) $27 + (18 \div 3) =$

6) $(18 \times 8) \div 6 =$

7) $(32 \div 4) \times (-2) =$

8) $(9 \times 4) + (32 - 18) =$

9) $24 + (4 \times 3) + 7 =$

10) $(36 \times 3) \div (2 + 2) =$

11) $(-7) + (12 \times 3) + 11 =$

12) $(8 \times 5) - (24 \div 6) =$

13) $(7 \times 6 \div 3) - (12 + 9) =$

14) $(13 + 5 - 14) \times 3 - 2 =$

15) $(20 - 14 + 30) \times (64 \div 4) =$

16) $32 + \left(28 - (36 \div 9)\right) =$

17) $(7 + 6 - 4 - 7) + (15 \div 5) =$

18) $(85 - 20) + (20 - 18 + 7) =$

19) $(20 \times 2) + (14 \times 3) - 22 =$

20) $18 + 5 - (30 \times 3) + 20 =$

Ordering Integers and Numbers

✎ **Order each set of integers from least to greatest.**

1) $8, -10, -5, -3, 4$

___, ___, ___, ___, ___, ___

2) $-10, -18, 6, 14, 27$

___, ___, ___, ___, ___, ___

3) $15, -8, -21, 21, -23$

___, ___, ___, ___, ___, ___

4) $-14, -40, 23, -12, 47$

___, ___, ___, ___, ___, ___

5) $59, -54, 32, -57, 36$

___, ___, ___, ___, ___, ___

6) $68, 26, -19, 47, -34$

___, ___, ___, ___, ___, ___

✎ **Order each set of integers from greatest to least.**

7) $18, 36, -16, -18, -10$

___, ___, ___, ___, ___, ___

8) $27, 34, -12, -24, 94$

___, ___, ___, ___, ___, ___

9) $50, -21, -13, 42, -2$

___, ___, ___, ___, ___, ___

10) $37, 46, -20, -16, 86$

___, ___, ___, ___, ___, ___

11) $-18, 88, -26, -59, 75$

___, ___, ___, ___, ___, ___

12) $-65, -30, -25, 3, 14$

___, ___, ___, ___, ___, ___

Integers and Absolute Value

✎ **Write absolute value of each number.**

1) $|-2| =$

2) $|-27| =$

3) $|-20| =$

4) $|14| =$

5) $|6| =$

6) $|-55| =$

7) $|16| =$

8) $|2| =$

9) $|54| =$

10) $|-4| =$

11) $|-11|$

12) $|88| =$

13) $|0| =$

14) $|79| =$

15) $|-32| =$

16) $|-17| =$

17) $|42| =$

18) $|-46| =$

19) $|1| =$

20) $|-40| =$

✎ **Evaluate the value.**

21) $|-5| - \dfrac{|-21|}{7} =$

22) $14 - |3 - 15| - |-4| =$

23) $\dfrac{|-32|}{4} \times |-4| =$

24) $\dfrac{|7 \times (-3)|}{7} \times \dfrac{|-19|}{3} =$

25) $|4 \times (-5)| + \dfrac{|-40|}{5} =$

26) $\dfrac{|-45|}{9} \times \dfrac{|-24|}{12} =$

27) $|-12 + 8| \times \dfrac{|-7 \times 7|}{7} =$

28) $\dfrac{|-11 \times 2|}{4} \times |-16| =$

Factoring Numbers

✎ **List all positive factors of each number.**

1) 9	9) 28	17) 50
2) 16	10) 98	18) 62
3) 24	11) 14	19) 95
4) 30	12) 54	20) 64
5) 26	13) 55	21) 70
6) 46	14) 18	22) 45
7) 20	15) 63	23) 22
8) 68	16) 34	24) 65

Greatest Common Factor

✎ **Find the GCF for each number pair.**

1) 6, 2   9) 12, 18   17) 42, 14

2) 4, 5   10) 4, 36   18) 16, 40

3) 3, 12   11) 6, 10   19) 9, 2, 3

4) 7, 3   12) 28, 52   20) 5, 15, 10

5) 5, 10   13) 25, 10   21) 7, 9, 2

6) 8, 48   14) 22, 24   22) 16, 64

7) 6, 18   15) 9, 54   23) 30, 48

8) 9, 15   16) 8, 54   24) 36, 63

Least Common Multiple

✍ **Find the LCM for each number pair.**

1) 6, 9

2) 15, 45

3) 16, 40

4) 12, 36

5) 18, 27

6) 14, 42

7) 6, 30

8) 8, 56

9) 7, 21

10) 8, 20

11) 15, 25

12) 7, 9

13) 4, 11

14) 8, 28

15) 28, 56

16) 40, 50

17) 12, 13

18) 22, 11

19) 36, 20

20) 15, 35

21) 18, 81

22) 30, 54

23) 18, 45

24) 75, 25

Sets

✍ Given A = {3, 4, 5, 7, 13}, B = {1, 6, 9, 10}, and C = {4, 6, 7, 10}, find:

1) A ∪ C _____   6) A ∩ C _____

2) A ∪ B _____   7) (C ∪ B) ∪ A_____

3) B ∪ C _____   8) (A ∪ C) ∩ B_____

4) C ∩ B _____   9) (C ∩ B) ∩ A_____

5) A ∩ B _____   10) (B ∪ A) ∩ C_____

✍ **Refer to the diagram below to find each set.**

11) A ∪ B _____

12) A ∪ C _____

13) C ∪ B _____

14) A ∩ C _____

15) B ∩ C _____

16) B ∩ A _____

17) (A ∪ C) ∪ B _____

18) (C ∪ B) ∩ A _____

19) (A ∩ B) ∩ C _____

20) (A ∪ C) ∩ B _____

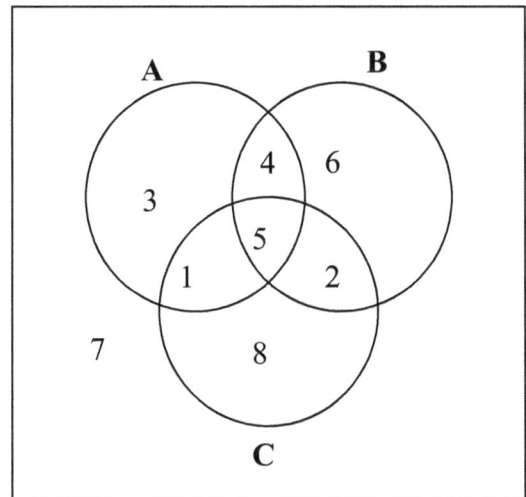

TSI Subject Test Mathematics

Answers of Worksheets

Rounding

1) 40	10) 80	19) 600	28) 7,000
2) 90	11) 60	20) 800	29) 9,000
3) 20	12) 90	21) 600	30) 23,000
4) 60	13) 200	22) 900	31) 45,000
5) 20	14) 400	23) 500	32) 17,000
6) 30	15) 800	24) 300	33) 53,000
7) 90	16) 100	25) 1,000	34) 85,000
8) 70	17) 300	26) 3,000	35) 71,000
9) 50	18) 400	27) 4,000	36) 27,000

Rounding and Estimates

1) 30	7) 760	13) 3,200	19) 130
2) 90	8) 4,610	14) 1,800	20) 110
3) 120	9) 2,100	15) 600	21) 400
4) 150	10) 300	16) 4,200	22) 100
5) 620	11) 2,400	17) 300	
6) 1,230	12) 200	18) 1,200	

Adding and Subtracting Integers

1) 8	8) 13	15) 42	22) 9
2) −33	9) 29	16) 36	23) 15
3) −23	10) 34	17) 101	24) 11
4) 38	11) −11	18) 40	25) 49
5) −19	12) 39	19) 32	26) 38
6) 24	13) −56	20) 112	
7) 33	14) 53	21) 40	

Multiplying and Dividing Integers

1) 45	6) −88	11) −6	16) 12
2) −27	7) 324	12) 8	17) −18
3) −96	8) −300	13) 5	18) 26
4) 140	9) −120	14) −27	19) −17
5) 90	10) −400	15) 8	20) −21

21) 23   23) 30   25) 26

22) 19   24) −15   26) −9

Order of Operations

1) 27	6) 24	11) 40	16) 56
2) −4	7) −16	12) 36	17) 5
3) 92	8) 50	13) −7	18) 74
4) −7	9) 43	14) 10	19) 60
5) 33	10) 27	15) 576	20) −47

Ordering Integers and Numbers

1) −10, −5, −3, 4, 8

2) −18, −10, 6, 14, 27

3) −23, −21, −8, 15, 21

4) −40, −14, −12, 23, 47

5) −57, −54, 32, 36, 59

6) −34, −19, 26, 47, 68

7) 36, 18, −10, −16, −18

8) 94, 34, 27, −12, −24

9) 50, 42, −2, −13, −21

10) 86, 46, 37, −16, −20

11) 88, 75, −18, −26, −59

12) 14, 3, −25, −30, −65

Integers and Absolute Value

1) 2	8) 2	15) 32	22) −2
2) 27	9) 54	16) 17	23) 32
3) 20	10) 4	17) 42	24) 19
4) 14	11) 11	18) 46	25) 28
5) 6	12) 88	19) 1	26) 10
6) 55	13) 0	20) 40	27) 28
7) 16	14) 79	21) 2	28) 88

Factoring Numbers

1) 1, 3, 9

2) 1, 2, 4, 8, 16

3) 1, 2, 3, 4, 6, 8, 12, 24

4) 1, 2, 3, 5, 6, 10, 15, 30

5) 1, 2, 13, 26

6) 1, 2, 23, 46

7) 1, 2, 4, 5, 10, 20

8) 1, 2, 4, 17, 34, 68

9) 1, 2, 4, 7, 14, 28

10) 1, 2, 7, 14, 49, 98

11) 1, 2, 7, 14

12) 1, 2, 3, 6, 9, 18, 27, 54

13) 1, 5, 11, 55

14) 1, 2, 3, 6, 9, 18

15) 1, 3, 7, 9, 21, 63

16) 1, 2, 17, 34

17) 1, 2, 5, 10, 25, 50

18) 1, 2, 31, 62

19) 1, 5, 19, 95

20) 1, 2, 4, 8, 16, 32, 64

21) 1, 2, 5, 7, 10, 14, 35, 70

22) 1, 3, 5, 9, 15, 45   23) 1, 2, 11, 22   24) 1, 5, 13, 65

Greatest Common Factor

1) 2	7) 6	13) 5	19) 1
2) 1	8) 3	14) 2	20) 5
3) 3	9) 6	15) 9	21) 1
4) 1	10) 4	16) 2	22) 16
5) 5	11) 2	17) 14	23) 6
6) 8	12) 4	18) 8	24) 9

Least Common Multiple

1) 18	7) 30	13) 44	19) 180
2) 45	8) 56	14) 56	20) 105
3) 80	9) 21	15) 56	21) 162
4) 36	10) 40	16) 200	22) 270
5) 54	11) 75	17) 156	23) 90
6) 42	12) 63	18) 22	24) 75

Sets

1) {3, 4, 5, 6, 7, 10, 13}	8) {6, 10}	15) {2, 5}
2) {1, 3, 4, 5, 6, 7, 9, 10, 13}	9) { } (empty set)	16) {4, 5}
3) {1, 4, 6, 7, 9, 10}	10) {4, 6, 7, 10}	17) {1, 2, 3, 4, 5, 6, 8}
4) {6, 10}	11) {1, 2, 3, 4, 5, 6}	18) {1, 4, 5}
5) { } (empty set)	12) {1, 2, 3, 4, 5, 8}	19) {5}
6) {4, 7}	13) {1, 2, 4, 5, 6, 8}	20) {2, 4, 5}
7) {1, 3, 4, 5, 6, 7, 9, 10, 13}	14) {1, 5}	

Chapter 2 :

Fractions and Decimals

Topics that you will practice in this chapter:

- ✓ Simplifying Fractions
- ✓ Adding and Subtracting Fractions
- ✓ Multiplying and Dividing Fractions
- ✓ Adding and Subtract Mixed Numbers
- ✓ Multiplying and Dividing Mixed Numbers
- ✓ Adding and Subtracting Decimals
- ✓ Multiplying and Dividing Decimals
- ✓ Comparing Decimals
- ✓ Rounding Decimals

"A Man is like a fraction whose numerator is what he is and whose denominator is what he thinks of himself. The larger the denominator, the smaller the fraction." –Tolstoy

Simplifying Fractions

✎ **Simplify each fraction to its lowest terms.**

1) $\frac{5}{10} =$

2) $\frac{28}{35} =$

3) $\frac{27}{36} =$

4) $\frac{40}{80} =$

5) $\frac{14}{56} =$

6) $\frac{32}{48} =$

7) $\frac{52}{65} =$

8) $\frac{15}{60} =$

9) $\frac{80}{160} =$

10) $\frac{55}{77} =$

11) $\frac{28}{112} =$

12) $\frac{32}{64} =$

13) $\frac{63}{72} =$

14) $\frac{81}{90} =$

15) $\frac{35}{105} =$

16) $\frac{25}{70} =$

17) $\frac{80}{280} =$

18) $\frac{12}{81} =$

19) $\frac{36}{186} =$

20) $\frac{240}{540} =$

21) $\frac{70}{560} =$

✎ **Find the answer for each problem.**

22) Which of the following fractions equal to $\frac{3}{4}$? ____

   A. $\frac{60}{90}$ B. $\frac{43}{104}$ C. $\frac{48}{64}$ D. $\frac{150}{300}$

23) Which of the following fractions equal to $\frac{5}{8}$? ____

   A. $\frac{125}{200}$ B. $\frac{115}{200}$ C. $\frac{50}{100}$ D. $\frac{30}{90}$

24) Which of the following fractions equal to $\frac{3}{7}$? ____

   A. $\frac{58}{116}$ B. $\frac{54}{126}$ C. $\frac{270}{167}$ D. $\frac{42}{63}$

Adding and Subtracting Fractions

✎ Find the sum.

1) $\frac{5}{9} + \frac{4}{9} =$

2) $\frac{1}{2} + \frac{1}{7} =$

3) $\frac{3}{8} + \frac{1}{4} =$

4) $\frac{3}{5} + \frac{1}{2} =$

5) $\frac{1}{4} + \frac{3}{5} =$

6) $\frac{7}{8} + \frac{3}{8} =$

7) $\frac{1}{2} + \frac{7}{10} =$

8) $\frac{2}{5} + \frac{2}{3} =$

9) $\frac{5}{7} + \frac{2}{3} =$

10) $\frac{7}{12} + \frac{3}{4} =$

11) $\frac{5}{6} + \frac{2}{5} =$

12) $\frac{1}{12} + \frac{2}{3} =$

✎ Find the difference.

13) $\frac{1}{3} - \frac{1}{6} =$

14) $\frac{3}{4} - \frac{1}{8} =$

15) $\frac{1}{2} - \frac{1}{3} =$

16) $\frac{1}{4} - \frac{1}{5} =$

17) $\frac{5}{8} - \frac{2}{3} =$

18) $\frac{1}{4} - \frac{1}{7} =$

19) $\frac{5}{6} - \frac{1}{9} =$

20) $\frac{3}{4} - \frac{1}{6} =$

21) $\frac{7}{8} - \frac{1}{12} =$

22) $\frac{8}{15} - \frac{3}{5} =$

23) $\frac{3}{12} - \frac{1}{14} =$

24) $\frac{10}{13} - \frac{7}{26} =$

25) $\frac{6}{7} - \frac{3}{4} =$

26) $\frac{4}{5} - \frac{1}{8} =$

27) $\frac{4}{7} - \frac{2}{35} =$

28) $\frac{9}{16} - \frac{2}{8} =$

29) $\frac{8}{9} - \frac{7}{18} =$

30) $\frac{1}{2} - \frac{4}{9} =$

Multiplying and Dividing Fractions

✏ **Find the value of each expression in lowest terms.**

1) $\frac{1}{5} \times \frac{15}{5} =$

2) $\frac{9}{12} \times \frac{4}{9} =$

3) $\frac{1}{16} \times \frac{8}{10} =$

4) $\frac{1}{24} \times \frac{8}{10} =$

5) $\frac{1}{5} \times \frac{1}{4} =$

6) $\frac{7}{9} \times \frac{1}{7} =$

7) $\frac{6}{7} \times \frac{1}{3} =$

8) $\frac{2}{8} \times \frac{2}{8} =$

9) $\frac{5}{8} \times \frac{3}{5} =$

10) $\frac{4}{7} \times \frac{1}{8} =$

11) $\frac{7}{15} \times \frac{5}{7} =$

12) $\frac{3}{10} \times \frac{5}{9} =$

✏ **Find the value of each expression in lowest terms.**

13) $\frac{1}{4} \div \frac{1}{8} =$

14) $\frac{1}{10} \div \frac{1}{5} =$

15) $\frac{3}{4} \div \frac{1}{5} =$

16) $\frac{1}{3} \div \frac{5}{6} =$

17) $\frac{1}{7} \div \frac{8}{42} =$

18) $\frac{3}{4} \div \frac{1}{6} =$

19) $\frac{2}{7} \div \frac{7}{13} =$

20) $\frac{1}{24} \div \frac{3}{16} =$

21) $\frac{7}{12} \div \frac{5}{6} =$

22) $\frac{22}{18} \div \frac{11}{9} =$

23) $\frac{9}{35} \div \frac{3}{7} =$

24) $\frac{2}{7} \div \frac{8}{21} =$

25) $\frac{1}{9} \div \frac{2}{5} =$

26) $\frac{5}{12} \div \frac{3}{5} =$

27) $\frac{3}{20} \div \frac{1}{6} =$

28) $\frac{8}{20} \div \frac{3}{4} =$

29) $\frac{5}{6} \div \frac{2}{9} =$

30) $\frac{5}{11} \div \frac{3}{4} =$

Adding and Subtracting Mixed Numbers

✎ **Find the sum.**

1) $3\frac{1}{3} + 2\frac{1}{6} =$

2) $4\frac{1}{2} + 3\frac{1}{2} =$

3) $3\frac{3}{8} + 1\frac{1}{8} =$

4) $2\frac{1}{4} + 2\frac{1}{3} =$

5) $3\frac{5}{6} + 2\frac{7}{12} =$

6) $5\frac{4}{15} + 3\frac{3}{5} =$

7) $2\frac{1}{3} + 4\frac{3}{7} =$

8) $3\frac{1}{2} + 4\frac{2}{5} =$

9) $5\frac{2}{5} + 6\frac{3}{7} =$

10) $8\frac{5}{16} + 6\frac{1}{12} =$

✎ **Find the difference.**

11) $3\frac{1}{4} - 1\frac{3}{4} =$

12) $6\frac{3}{5} - 4\frac{2}{5} =$

13) $4\frac{1}{3} - 3\frac{1}{9} =$

14) $7\frac{1}{7} - 5\frac{1}{2} =$

15) $5\frac{1}{3} - 2\frac{1}{12} =$

16) $8\frac{1}{5} - 4\frac{1}{3} =$

17) $9\frac{1}{4} - 6\frac{1}{8} =$

18) $11\frac{7}{15} - 8\frac{3}{5} =$

19) $14\frac{5}{6} - 11\frac{3}{5} =$

20) $18\frac{2}{7} - 14\frac{1}{5} =$

21) $9\frac{1}{3} - 4\frac{1}{4} =$

22) $6\frac{1}{8} - 4\frac{1}{16} =$

23) $19\frac{3}{8} - 15\frac{1}{3} =$

24) $11\frac{1}{9} - 8\frac{1}{8} =$

25) $17\frac{1}{7} - 11\frac{1}{5} =$

26) $16\frac{2}{9} - 9\frac{5}{7} =$

Multiplying and Dividing Mixed Numbers

✎ **Find the product.**

1) $5\frac{1}{2} \times 2\frac{1}{4} =$

2) $5\frac{1}{3} \times 4\frac{1}{3} =$

3) $5\frac{3}{4} \times 6\frac{1}{4} =$

4) $3\frac{1}{3} \times 2\frac{3}{5} =$

5) $4\frac{8}{10} \times 1\frac{1}{24} =$

6) $6\frac{2}{7} \times 1\frac{1}{11} =$

7) $8\frac{2}{3} \times 3\frac{1}{2} =$

8) $3\frac{4}{7} \times 2\frac{1}{5} =$

9) $5\frac{2}{8} \times 4\frac{1}{6} =$

10) $7\frac{3}{3} \times 1\frac{3}{8} =$

✎ **Find the quotient.**

11) $2\frac{2}{5} \div 4\frac{1}{5} =$

12) $4\frac{1}{6} \div 3\frac{1}{3} =$

13) $6\frac{1}{3} \div 1\frac{1}{2} =$

14) $7\frac{1}{10} \div 2\frac{2}{5} =$

15) $3\frac{1}{3} \div 1\frac{1}{9} =$

16) $1\frac{1}{10} \div 4\frac{1}{2} =$

17) $1\frac{3}{16} \div 5\frac{1}{4} =$

18) $4\frac{1}{3} \div 4\frac{3}{4} =$

19) $9\frac{1}{3} \div 2\frac{1}{4} =$

20) $15\frac{1}{3} \div 5\frac{1}{2} =$

21) $4\frac{1}{6} \div 1\frac{1}{5} =$

22) $1\frac{1}{18} \div 1\frac{2}{9} =$

23) $4\frac{2}{7} \div 1\frac{3}{10} =$

24) $7\frac{1}{3} \div 2\frac{2}{11} =$

25) $8\frac{2}{5} \div 1\frac{1}{6} =$

26) $9\frac{1}{3} \div 2\frac{1}{7} =$

Adding and Subtracting Decimals

✎ **Add and subtract decimals.**

1)
$$\begin{array}{r} 35.19 \\ -\ 24.28 \\ \hline \end{array}$$

4)
$$\begin{array}{r} 38.72 \\ -\ 21.68 \\ \hline \end{array}$$

7)
$$\begin{array}{r} 86.09 \\ -\ 35.14 \\ \hline \end{array}$$

2)
$$\begin{array}{r} 34.29 \\ +\ 42.58 \\ \hline \end{array}$$

5)
$$\begin{array}{r} 57.39 \\ +\ 26.54 \\ \hline \end{array}$$

8)
$$\begin{array}{r} 54.51 \\ +\ 32.66 \\ \hline \end{array}$$

3)
$$\begin{array}{r} 61.20 \\ +\ 33.75 \\ \hline \end{array}$$

6)
$$\begin{array}{r} 70.24 \\ -\ 42.35 \\ \hline \end{array}$$

9)
$$\begin{array}{r} 114.21 \\ -\ 88.69 \\ \hline \end{array}$$

✎ **Find the missing number.**

10) ___ $+ 2.8 = 5.4$

11) $4.1 +$ ___ $= 5.88$

12) $6.45 +$ ___ $= 8$

13) $7.25 -$ ___ $= 3.40$

14) ___ $- 2.35 = 4.25$

15) ___ $- 19.85 = 6.54$

16) $22.15 +$ ___ $= 28.95$

17) ___ $- 37.16 = 9.42$

18) ___ $+ 24.50 = 34.19$

19) $72.40 +$ ___ $= 125.20$

Multiplying and Dividing Decimals

✎ **Find the product.**

1) $0.5 \times 0.6 =$

2) $3.3 \times 0.4 =$

3) $1.28 \times 0.5 =$

4) $0.35 \times 0.6 =$

5) $1.85 \times 0.6 =$

6) $0.24 \times 0.5 =$

7) $5.25 \times 1.4 =$

8) $18.5 \times 4.6 =$

9) $15.4 \times 6.8 =$

10) $19.5 \times 2.6 =$

11) $32.2 \times 1.5 =$

12) $78.4 \times 4.5 =$

✎ **Find the quotient.**

13) $1.85 \div 10 =$

14) $74.6 \div 100 =$

15) $3.6 \div 3 =$

16) $9.6 \div 0.4 =$

17) $15.5 \div 0.5 =$

18) $32.8 \div 0.2 =$

19) $22.15 \div 1,000 =$

20) $53.55 \div 0.7 =$

21) $322.2 \div 0.2 =$

22) $50.67 \div 0.18 =$

23) $77.4 \div 0.8 =$

24) $27.93 \div 0.03 =$

Comparing Decimals

✎ **Write the correct comparison symbol (>, < or =).**

1) 0.70 ☐ 0.070

2) 0.049 ☐ 0.49

3) 5.090 ☐ 5.09

4) 2.57 ☐ 2.05

5) 9.03 ☐ 0.930

6) 6.06 ☐ 6.6

7) 7.02 ☐ 7.020

8) 3.04 ☐ 3.2

9) 3.61 ☐ 3.245

10) 0.986 ☐ 0.0986

11) 17.24 ☐ 17.240

12) 0.759 ☐ 0.81

13) 9.040 ☐ 9.40

14) 5.73 ☐ 5.213

15) 9.44 ☐ 9.404

16) 7.17 ☐ 7.170

17) 4.85 ☐ 4.085

18) 9.041 ☐ 9.40

19) 3.033 ☐ 3.030

20) 4.97 ☐ 4.970

Rounding Decimals

✎ **Round each decimal to the nearest whole number.**

1) 28.12	3) 16.22	5) 7.95
2) 6.9	4) 8.5	6) 52.7

✎ **Round each decimal to the nearest tenth.**

7) 31.761	9) 94.729	11) 13.219
8) 14.421	10) 77.89	12) 59.89

✎ **Round each decimal to the nearest hundredth.**

13) 8.428	15) 55.3786	17) 62.241
14) 23.812	16) 231.912	18) 19.447

✎ **Round each decimal to the nearest thousandth.**

19) 15.54324	21) 243.8652	23) 67.1983
20) 34.62586	22) 80.4529	24) 72.36788

Answers of Worksheets

Simplifying Fractions

1) $\frac{1}{2}$

2) $\frac{4}{5}$

3) $\frac{3}{4}$

4) $\frac{1}{2}$

5) $\frac{1}{4}$

6) $\frac{2}{3}$

7) $\frac{4}{5}$

8) $\frac{1}{4}$

9) $\frac{1}{2}$

10) $\frac{5}{7}$

11) $\frac{1}{4}$

12) $\frac{1}{2}$

13) $\frac{7}{8}$

14) $\frac{9}{10}$

15) $\frac{1}{3}$

16) $\frac{5}{14}$

17) $\frac{2}{7}$

18) $\frac{4}{27}$

19) $\frac{6}{31}$

20) $\frac{4}{9}$

21) $\frac{1}{8}$

22) C

23) A

24) B

Adding and Subtracting Fractions

1) $\frac{9}{9} = 1$

2) $\frac{9}{14}$

3) $\frac{5}{8}$

4) $1\frac{1}{10}$

5) $\frac{17}{20}$

6) $1\frac{1}{4}$

7) $1\frac{1}{5}$

8) $1\frac{1}{15}$

9) $1\frac{8}{21}$

10) $1\frac{1}{3}$

11) $1\frac{7}{30}$

12) $\frac{3}{4}$

13) $\frac{1}{6}$

14) $\frac{5}{8}$

15) $\frac{1}{6}$

16) $\frac{1}{20}$

17) $-\frac{1}{24}$

18) $\frac{3}{28}$

19) $\frac{13}{18}$

20) $\frac{7}{12}$

21) $\frac{19}{24}$

22) $-\frac{1}{15}$

23) $\frac{5}{28}$

24) $\frac{1}{2}$

25) $\frac{3}{28}$

26) $\frac{27}{40}$

27) $\frac{18}{35}$

28) $\frac{5}{16}$

29) $\frac{1}{2}$

30) $\frac{1}{18}$

Multiplying and Dividing Fractions

1) $\frac{3}{5}$

2) $\frac{1}{3}$

3) $\frac{1}{20}$

4) $\frac{1}{30}$

5) $\frac{1}{20}$

6) $\frac{1}{9}$

7) $\frac{2}{7}$

8) $\frac{1}{16}$

9) $\frac{3}{8}$

10) $\frac{1}{14}$

11) $\frac{1}{3}$

12) $\frac{1}{6}$

13) 2

14) $\frac{1}{2}$

15) $3\frac{3}{4}$

16) $\frac{2}{5}$

17) $\frac{3}{4}$

18) $4\frac{1}{2}$

19) $\frac{26}{49}$

20) $\frac{2}{9}$

21) $\frac{7}{10}$

22) 1

23) $\frac{3}{5}$

24) $\frac{3}{4}$

25) $\frac{5}{18}$

26) $\frac{25}{36}$

27) $\frac{9}{10}$

28) $\frac{8}{15}$

29) $3\frac{3}{4}$

30) $\frac{20}{33}$

Adding and Subtracting Mixed Numbers

1) $5\frac{1}{2}$

2) 8

3) $4\frac{1}{2}$

4) $4\frac{7}{12}$

5) $6\frac{5}{12}$

6) $8\frac{13}{15}$

7) $6\frac{16}{21}$

8) $7\frac{9}{10}$

9) $11\frac{29}{35}$

10) $14\frac{19}{48}$

11) $1\frac{1}{2}$

12) $2\frac{1}{5}$

13) $1\frac{2}{9}$

14) $1\frac{9}{14}$

15) $3\frac{1}{4}$

16) $3\frac{13}{15}$

17) $3\frac{1}{8}$

18) $2\frac{13}{15}$

19) $3\frac{7}{30}$

20) $4\frac{3}{35}$

21) $5\frac{1}{12}$

22) $2\frac{1}{16}$

23) $4\frac{1}{24}$

24) $2\frac{71}{72}$

25) $5\frac{33}{35}$

26) $6\frac{32}{63}$

Multiplying and Dividing Mixed Numbers

1) $12\frac{3}{8}$

2) $23\frac{1}{9}$

3) $35\frac{15}{16}$

4) $8\frac{2}{3}$

5) 5

6) $6\frac{6}{7}$

7) $30\frac{1}{3}$

8) $7\frac{6}{7}$

9) $21\frac{7}{8}$

10) 11

11) $\frac{4}{7}$

12) $1\frac{1}{4}$

13) $4\frac{2}{9}$

14) $2\frac{23}{24}$

15) 3

16) $\frac{11}{45}$

17) $\frac{19}{84}$

18) $\frac{52}{57}$

19) $4\frac{4}{27}$

20) $2\frac{26}{33}$

21) $3\frac{17}{36}$

22) $\frac{19}{22}$

23) $3\frac{27}{91}$

24) $3\frac{13}{36}$

25) $7\frac{1}{5}$

26) $4\frac{16}{45}$

Adding and Subtracting Decimals

1) 10.91

2) 76.87

3) 94.95

4) 17.04

5) 83.93

6) 27.89

7) 50.95

8) 87.17

9) 25.52

10) 2.6

11) 1.78

12) 1.55

13) 3.85

14) 6.6

15) 26.39

16) 6.8

17) 46.58

18) 9.69

19) 52.8

TSI Subject Test Mathematics

Multiplying and Dividing Decimals

1) 0.3	7) 7.35	13) 0.185	19) 0.02215
2) 1.32	8) 85.1	14) 0.746	20) 76.5
3) 0.64	9) 104.72	15) 1.2	21) 1,611
4) 0.21	10) 50.7	16) 24	22) 281.5
5) 1.11	11) 48.3	17) 31	23) 96.75
6) 0.12	12) 352.8	18) 164	24) 931

Comparing Decimals

1) >	6) <	11) =	16) =
2) <	7) =	12) <	17) >
3) =	8) <	13) <	18) <
4) >	9) >	14) >	19) >
5) >	10) >	15) >	20) =

Rounding Decimals

1) 28	9) 94.7	17) 62.24
2) 7	10) 77.9	18) 19.45
3) 16	11) 13.2	19) 15.543
4) 9	12) 59.9	20) 34.626
5) 8	13) 8.43	21) 243.865
6) 53	14) 23.81	22) 80.453
7) 31.8	15) 55.38	23) 67.198
8) 14.4	16) 231.91	24) 72.368

Chapter 3 :

Proportions, Ratios, and Percent

Topics that you will practice in this chapter:

✓ Simplifying Ratios

✓ Proportional Ratios

✓ Similarity and Ratios

✓ Ratio and Rates Word Problems

✓ Percentage Calculations

✓ Percent Problems

✓ Discount, Tax and Tip

✓ Percent of Change

✓ Simple Interest

Without mathematics, there's nothing you can do. Everything around you is mathematics. Everything around you is numbers." – Shakuntala Devi

Simplifying Ratios

✍ **Reduce each ratio.**

1) $15:20 = \underline{\quad}:\underline{\quad}$

2) $7:70 = \underline{\quad}:\underline{\quad}$

3) $16:28 = \underline{\quad}:\underline{\quad}$

4) $7:21 = \underline{\quad}:\underline{\quad}$

5) $4:40 = \underline{\quad}:\underline{\quad}$

6) $6:48 = \underline{\quad}:\underline{\quad}$

7) $16:64 = \underline{\quad}:\underline{\quad}$

8) $10:25 = \underline{\quad}:\underline{\quad}$

9) $8:48 = \underline{\quad}:\underline{\quad}$

10) $49:63 = \underline{\quad}:\underline{\quad}$

11) $18:27 = \underline{\quad}:\underline{\quad}$

12) $35:10 = \underline{\quad}:\underline{\quad}$

13) $90:9 = \underline{\quad}:\underline{\quad}$

14) $24:32 = \underline{\quad}:\underline{\quad}$

15) $7:56 = \underline{\quad}:\underline{\quad}$

16) $45:63 = \underline{\quad}:\underline{\quad}$

17) $56:72 = \underline{\quad}:\underline{\quad}$

18) $26:13 = \underline{\quad}:\underline{\quad}$

19) $15:45 = \underline{\quad}:\underline{\quad}$

20) $28:4 = \underline{\quad}:\underline{\quad}$

21) $24:48 = \underline{\quad}:\underline{\quad}$

22) $30:24 = \underline{\quad}:\underline{\quad}$

23) $70:140 = \underline{\quad}:\underline{\quad}$

24) $6:180 = \underline{\quad}:\underline{\quad}$

✍ **Write each ratio as a fraction in simplest form.**

25) $6:12 =$

26) $30:50 =$

27) $15:35 =$

28) $9:27 =$

29) $8:24 =$

30) $18:84 =$

31) $7:14 =$

32) $7:35 =$

33) $40:96 =$

34) $12:54 =$

35) $44:52 =$

36) $12:27 =$

37) $15:180 =$

38) $39:143 =$

39) $20:300 =$

40) $30:120 =$

41) $56:42 =$

42) $26:130 =$

43) $66:123 =$

44) $70:630 =$

45) $75:125 =$

Proportional Ratios

✎ **Fill in the blanks; Calculate each proportion.**

1) $3:8 = \underline{\quad}:48$ 7) $20:3 = \underline{\quad}:15$

2) $2:5 = 20:\underline{\quad}$ 8) $1:3 = \underline{\quad}:75$

3) $1:9 = \underline{\quad}:81$ 9) $7:6 = \underline{\quad}:60$

4) $6:7 = 12:\underline{\quad}$ 10) $8:5 = \underline{\quad}:45$

5) $9:2 = 63:\underline{\quad}$ 11) $3:10 = 60:\underline{\quad}$

6) $8:7 = \underline{\quad}:49$ 12) $6:11 = 42:\underline{\quad}$

✎ **State if each pair of ratios form a proportion.**

13) $\frac{3}{20}$ and $\frac{9}{60}$ 17) $\frac{1}{9}$ and $\frac{12}{81}$ 21) $\frac{6}{19}$ and $\frac{30}{85}$

14) $\frac{1}{7}$ and $\frac{6}{42}$ 18) $\frac{7}{8}$ and $\frac{21}{28}$ 22) $\frac{5}{9}$ and $\frac{40}{81}$

15) $\frac{3}{7}$ and $\frac{24}{56}$ 19) $\frac{9}{13}$ and $\frac{27}{39}$ 23) $\frac{9}{14}$ and $\frac{108}{168}$

16) $\frac{4}{9}$ and $\frac{12}{18}$ 20) $\frac{1}{8}$ and $\frac{8}{64}$ 24) $\frac{15}{23}$ and $\frac{360}{552}$

✎ **Calculate each proportion.**

25) $\frac{20}{25} = \frac{32}{x}, x = \underline{\quad}$ 29) $\frac{7}{9} = \frac{x}{81}, x = \underline{\quad}$ 33) $\frac{5}{8} = \frac{x}{88}, x = \underline{\quad}$

26) $\frac{1}{8} = \frac{32}{x}, x = \underline{\quad}$ 30) $\frac{1}{5} = \frac{13}{x}, x = \underline{\quad}$ 34) $\frac{4}{15} = \frac{x}{240}, x = \underline{\quad}$

27) $\frac{15}{5} = \frac{21}{x}, x = \underline{\quad}$ 31) $\frac{9}{5} = \frac{36}{x}, x = \underline{\quad}$ 35) $\frac{9}{19} = \frac{x}{266}, x = \underline{\quad}$

28) $\frac{1}{7} = \frac{x}{294}, x = \underline{\quad}$ 32) $\frac{6}{13} = \frac{48}{x}, x = \underline{\quad}$ 36) $\frac{7}{15} = \frac{x}{270}, x = \underline{\quad}$

Similarity and Ratios

✎ **Each pair of figures is similar. Find the missing side.**

1)

2)

3)

4)

✎ **Calculate.**

5) Two rectangles are similar. The first is 24 feet wide and 120 feet long. The second is 30 feet wide. What is the length of the second rectangle? _____

6) Two rectangles are similar. One is 5 meters by 36 meters. The longer side of the second rectangle is 90 meters. What is the other side of the second rectangle? _____

7) A building casts a shadow 25 ft long. At the same time a girl 10 ft tall casts a shadow 5 ft long. How tall is the building? _____

8) The scale of a map of Texas is 4 inches: 32 miles. If you measure the distance from Dallas to Martin County as 38.4 inches, approximately how far is Martin County from Dallas? _____

Ratio and Rates Word Problems

✍ **Find the answer for each word problem.**

1) Mason has 24 red cards and 36 green cards. What is the ratio of Mason 's red cards to his green cards? _____

2) In a party, 45 soft drinks are required for every 54 guests. If there are 378 guests, how many soft drinks is required? _____

3) In Mason's class, 42 of the students are tall and 24 are short. In Michael's class 84 students are tall and 48 students are short. Which class has a higher ratio of tall to short students? _____

4) The price of 5 apples at the Quick Market is $4.6. The price of 7 of the same apples at Walmart is $5.95. Which place is the better buy? _____

5) The bakers at a Bakery can make 90 bagels in 3 hours. How many bagels can they bake in 24 hours? What is that rate per hour? _____

6) You can buy 5 cans of green beans at a supermarket for $5.75. How much does it cost to buy 45 cans of green beans? _____

7) The ratio of boys to girls in a class is 4: 7. If there are 32 boys in the class, how many girls are in that class? _____

8) The ratio of red marbles to blue marbles in a bag is 3: 7. If there are 50 marbles in the bag, how many of the marbles are red? _____

Percentage Calculations

✎ **Calculate the given percent of each value.**

1) 3% of 60 = ___

2) 20% of 32 = ___

3) 4% of 72 = ___

4) 16% of 32 = ___

5) 25% of 124 = ___

6) 35% of 56 = ___

7) 15% of 20 = ___

8) 14% of 150 = ___

9) 80% of 50 = ___

10) 12% of 115 = ___

11) 72% of 250 = ___

12) 52% of 500 = ___

13) 70% of 400 = ___

14) 27% of 145 = ___

15) 90% of 64 = ___

16) 60% of 55 = ___

17) 22% of 210 = ___

18) 8% of 235 = ___

✎ **Calculate the percent of each given value.**

19) ___% of 25 = 5

20) ___% of 40 = 20

21) ___% of 25 = 2

22) ___% of 50 = 16

23) ___% of 250 = 5

24) ___% of 40 = 32

25) ___% of 125 = 20

26) ___% of 700 = 49

27) ___% of 350 = 49

28) ___% of 500 = 210

✎ **Calculate each percent problem.**

29) A Cinema has 250 seats. 60 seats were sold for the current movie. What percent of seats are empty? ___ %

30) There are 68 boys and 92 girls in a class. 75% of the students in the class take the bus to school. How many students do not take the bus to school? ___

Percent Problems

✎ **Calculate each problem.**

1) 9 is what percent of 45? ____%

2) 60 is what percent of 120? ____%

3) 10 is what percent of 200? ____%

4) 15 is what percent of 125? ____%

5) 10 is what percent of 400? ____%

6) 66 is what percent of 55? ____%

7) 40 is what percent of 160? ____%

8) 40 is what percent of 50? ____%

9) 120 is what percent of 800? ____%

10) 78 is what percent of 120? ___%

11) 36 is what percent of 144? ___%

12) 17 is what percent of 85? ___%

13) 90 is what percent of 900? ___%

14) 36 is what percent of 16? ___%

15) 63 is what percent of 14? ___%

16) 18 is what percent of 60? ___%

17) 126 is what percent of 200? ___%

18) 232 is what percent of 40? ___%

✎ **Calculate each percent word problem.**

19) There are 40 employees in a company. On a certain day, 25 were present. What percent showed up for work? ____%

20) A metal bar weighs 60 ounces. 25% of the bar is gold. How many ounces of gold are in the bar? _____

21) A crew is made up of 12 women; the rest are men. If 15% of the crew are women, how many people are in the crew? _____

22) There are 40 students in a class and 8 of them are girls. What percent are boys? ____%

23) The Royals softball team played 400 games and won 280 of them. What percent of the games did they lose? ____%

Discount, Tax and Tip

✍ **Find the selling price of each item.**

1) Original price of a computer: $420

 Tax: 8% Selling price: $_____

2) Original price of a laptop: $280

 Tax: 4% Selling price: $_____

3) Original price of a sofa: $820

 Tax: 5% Selling price: $_____

4) Original price of a car: $15,800

 Tax: 3.6% Selling price: $_____

5) Original price of a Table: $250

 Tax: 9% Selling price: $_____

6) Original price of a house: $630,000

 Tax: 1.8% Selling price: $_____

7) Original price of a tablet: $450

 Discount: 30% Selling price: $____

8) Original price of a chair: $390

 Discount: 8% Selling price: $____

9) Original price of a book: $75

 Discount: 42% Selling price: $____

10) Original price of a cellphone: $820

 Discount: 23% Selling price: $___

11) Food bill: $45

 Tip: 15% Price: $_____

12) Food bill: $32

 Tipp: 20% Price: $_____

13) Food bill: $90

 Tip: 35% Price: $_____

14) Food bill: $42

 Tipp: 12% Price: $_____

✍ **Find the answer for each word problem.**

15) Nicolas hired a moving company. The company charged $500 for its services, and Nicolas gives the movers a 40% tip. How much does Nicolas tip the movers? $_____

16) Mason has lunch at a restaurant and the cost of his meal is $90. Mason wants to leave a 25% tip. What is Mason's total bill including tip? $_____

17) The sales tax in Texas is 19.80% and an item costs $350. How much is the tax? $_____

18) The price of a table at Best Buy is $680. If the sales tax is 5%, what is the final price of the table including tax? $_____

Percent of Change

✍ **Find each percent of change.**

1) From 150 to 450. ___ %

2) From 50 ft to 250 ft. ___ %

3) From $60 to $360. ___ %

4) From 60 cm to 180 cm. ___ %

5) From 15 to 45. ___ %

6) From 80 to 16.   ___ %

7) From 120 to 360. ___ %

8) From 900 to 450. ___ %

9) From 1,000 to 200.   ___ %

10) From 144 to 36.   ___ %

✍ **Calculate each percent of change word problem.**

11) Bob got a raise, and his hourly wage increased from $42 to $63. What is the percent increase? ___ %

12) The price of a pair of shoes increases from $50 to $61. What is the percent increase? ___ %

13) At a coffee shop, the price of a cup of coffee increased from $4.80 to $5.76. What is the percent increase in the cost of the coffee? ___ %

14) 51 cm are cut from 85 cm board. What is the percent decrease in length? ___ %

15) In a class, the number of students has been increased from 54 to 81. What is the percent increase? ___ %

16) The price of gasoline rises from $24.40 to $30.50 in one month. By what percent did the gas price rise? ___ %

17) A shirt was originally priced at $38. It went on sale for $24.70. What was the percent that the shirt was discounted? ___ %

Simple Interest

✍ Determine the simple interest for these loans.

1) $480 at 11% for 3 years. $ _____

2) $4,200 at 7% for 4 years. $ _____

3) $2,500 at 20% for 3 years. $ _____

4) $6,800 at 3.9% for 4 months. $ ____

5) $800 at 6% for 7 months. $ _____

6) $36,000 at 4.2% for 6 years. $ _____

7) $6,500 at 7% for 4 years. $ _____

8) $850 at 9.5% for 2 years. $ _____

9) $1,200 at 5.8% for 9 months. $ ____

10) $3,000 at 4.5% for 7 years. $ _____

✍ Calculate each simple interest word problem.

11) A new car, valued at $22,000, depreciates at 8.5% per year. What is the value of the car one year after purchase? $_____

12) Sara puts $9,000 into an investment yielding 6% annual simple interest; she left the money in for three years. How much interest does Sara get at the end of those three years? $_____

13) A bank is offering 12% simple interest on a savings account. If you deposit $16,400, how much interest will you earn in two years? $_____

14) $720 interest is earned on a principal of $6,000 at a simple interest rate of 4% interest per year. For how many years was the principal invested? _____

15) In how many years will $2,200 yield an interest of $440 at 4% simple interest? _____

16) Jim invested $8,000 in a bond at a yearly rate of 4.5%. He earned $1,440 in interest. How long was the money invested? _____

Answers of Worksheets

Simplifying Ratios

1) 3:4
2) 1:10
3) 4:7
4) 1:3
5) 1:10
6) 1:8
7) 2:8
8) 2:5
9) 1:6
10) 7:9
11) 2:3
12) 7:2
13) 10:1

14) 3:4
15) 1:8
16) 5:7
17) 7:9
18) 2:1
19) 1:3
20) 7:1
21) 1:2
22) 5:4
23) 1:2
24) 1:30
25) $\frac{1}{2}$

26) $\frac{3}{5}$
27) $\frac{3}{7}$
28) $\frac{1}{3}$
29) $\frac{1}{3}$
30) $\frac{3}{14}$
31) $\frac{1}{2}$
32) $\frac{1}{5}$
33) $\frac{5}{12}$
34) $\frac{2}{9}$
35) $\frac{11}{13}$

36) $\frac{4}{9}$
37) $\frac{1}{12}$
38) $\frac{3}{11}$
39) $\frac{1}{15}$
40) $\frac{1}{4}$
41) $\frac{4}{3}$
42) $\frac{1}{5}$
43) $\frac{22}{41}$
44) $\frac{1}{9}$
45) $\frac{3}{5}$

Proportional Ratios

1) 18
2) 50
3) 9
4) 14
5) 14
6) 56
7) 100
8) 25
9) 70

10) 72
11) 200
12) 77
13) Yes
14) Yes
15) Yes
16) No
17) No
18) No

19) Yes
20) Yes
21) No
22) No
23) Yes
24) Yes
25) 40
26) 256
27) 7

28) 42
29) 63
30) 65
31) 20
32) 104
33) 55
34) 64
35) 126
36) 126

Similarity and ratios

1) 15
2) 5
3) 15
4) 13
5) 150 feet
6) 12.5 meters
7) 50 feet
8) 307.2 miles

Ratio and Rates Word Problems

1) 2:3
2) 315

3) The ratio for both classes is 7 to 4.   6) $51.75

4) Walmart is a better buy.   7) 56

5) 720, the rate is 30 per hour.   8) 15

Percentage Calculations

1) 1.8	11) 180	21) 8%
2) 6.4	12) 260	22) 32%
3) 2.88	13) 280	23) 2%
4) 5.12	14) 39.15	24) 80%
5) 31	15) 57.6	25) 16%
6) 19.6	16) 33	26) 7%
7) 3	17) 46.2	27) 14%
8) 21	18) 18.8	28) 42%
9) 40	19) 20%	29) 76%
10) 13.8	20) 50%	30) 40

Percent Problems

1) 20%	9) 15%	17) 63%
2) 50%	10) 65%	18) 580%
3) 5%	11) 25%	19) 62.5%
4) 12%	12) 20%	20) 15 ounces
5) 2.5%	13) 10%	21) 80
6) 120%	14) 225%	22) 80%
7) 25%	15) 450%	23) 30%
8) 80%	16) 30%	

Discount, Tax and Tip

1) $453.60	7) $315.00	13) $121.50
2) $291.20	8) $358.80	14) $47.04
3) $861.00	9) $43.50	15) $200.00
4) $16,368.80	10) $631.40	16) $112.50
5) $272.50	11) $51.75	17) $69.30
6) $641,340	12) $38.40	18) $714.00

Percent of Change

1) 200%	7) 200%	13) 20%
2) 400%	8) 50%	14) 60%
3) 500%	9) 80%	15) 50%
4) 200%	10) 75%	16) 25%
5) 200%	11) 50%	17) 35%
6) 80%	12) 22%	

Simple Interest

1) $158.40	7) $1,820.00	13) $3,936.00
2) $1,176.00	8) $161.50	14) 3 years
3) $1,500.00	9) $52.20	15) 5 years
4) $88.40	10) $945.00	16) 4 years
5) $28.00	11) $20,130.00	
6) $9,072.00	12) $1,620.00	

Chapter 4 :

Algebraic Expressions

Topics that you will practice in this chapter:

- ✓ Simplifying Variable Expressions
- ✓ Simplifying Polynomial Expressions
- ✓ Translate Phrases into an Algebraic Statement
- ✓ The Distributive Property
- ✓ Evaluating One Variable Expressions
- ✓ Evaluating Two Variables Expressions
- ✓ Combining like Terms

I want freedom for the full expression of my personality.
Mahatma Gandhi

Simplifying Variable Expressions

✎ **Simplify each expression.**

1) $3(x + 5) =$

2) $(-4)(7x - 5) =$

3) $11x + 5 - 6x =$

4) $-4 - 2x^2 - 6x^2 =$

5) $7 + 13x^2 + 3 =$

6) $3x^2 + 7x + 15x^2 =$

7) $3x^2 - 12x^2 + 4x =$

8) $4x^2 - 8x - 2x =$

9) $6x + 7(3 - 4x) =$

10) $8x + 4(15x - 3) =$

11) $6(-3x - 9) - 17 =$

12) $-11x^2 - (-5x) =$

13) $2x + 7 + 5 - 8x =$

14) $7 + 6x - 11 - 5x =$

15) $27x + 8 - 13 - 5x =$

16) $(-11)(-5x + 2) - 41x =$

17) $19x - 4(4 - 2x) =$

18) $16x + 3(3x + 6) + 10 =$

19) $5(-2x - 4) - 13x =$

20) $16x - 3x(x + 10) =$

21) $17x + 5x(2 - 4x) =$

22) $5x(-4x - 7) + 20x =$

23) $25x - 19 + 4x^2 =$

24) $6x(x - 11) + 25 =$

25) $4x - 5 + 15x + 3x^2 =$

26) $-7x^2 - 11x - 9x =$

27) $10x - 9x^2 - 3x^2 - 7 =$

28) $13 + 3x^2 - 9x^2 - 21x =$

29) $22x + 10x^2 - 15x + 17 =$

30) $4x^2 + 25x + 21x^2 =$

31) $29 - 12x^2 - 23x - 4x^2 =$

32) $22x - 19x - 9x^2 + 30 =$

Simplifying Polynomial Expressions

✍ **Simplify each polynomial.**

1) $(2x^3 + 8x^2) - (11x + 3x^2) =$ _____

2) $(2x^5 + 7x^3) - (5x^3 + 11x^2) =$ _____

3) $(41x^4 + 5x^2) - (4x^2 + 20x^4) =$ _____

4) $13x - 8x^2 + 4(4x^2 + 3x^3) =$ _____

5) $(4x^3 - 22) + 5(3x^2 - 6x^3) =$ _____

6) $(4x^3 - 3x) - 5(2x^3 + x^4) =$ _____

7) $5(5x - 2x^3) - 2(8x^3 + 5x^2) =$ _____

8) $(3x^2 - 10x) - (5x^3 + 14x^2) =$ _____

9) $5x^3 - (3x^4 + 5x) + 2x^2 =$ _____

10) $11x^4 - (3x^2 + 5x) + 7x =$ _____

11) $(6x^2 - 3x^4) - (10x^4 + 3x^2) =$ _____

12) $2x^2 - 7x^3 + 19x^4 - 22x^3 =$ _____

13) $10x^2 - x^4 + 4x^4 - 32x^3 =$ _____

14) $-5x^2 + 17x^3 - 8x^2 - 6x =$ _____

15) $x^4 - 11x^5 - 30x^4 + 5x^2 =$ _____

16) $21x^3 + 13x - 5x^2 - 11x^3 =$ _____

Translate Phrases into an Algebraic Statement

✏️ **Write an algebraic expression for each phrase.**

1) 9 multiplied by x. _____

2) Subtract 11 from y. _____

3) 19 divided by x. _____

4) 38 decreased by y. _____

5) Add y to 40. _____

6) The square of 6. _____

7) x raised to the fifth power. _____

8) The sum of six and a number. _____

9) The difference between fifty–seven and y. _____

10) The quotient of nine and a number. _____

11) The quotient of the square of x and 25. _____

12) The difference between x and 6 is 19. _____

13) 10 times a reduced by the square of b. _____

14) Subtract the product of a and b from 41. _____

The Distributive Property

✍ **Use the distributive property to simply each expression.**

1) $4(1 + 2x) =$

2) $2(4 + 7x) =$

3) $3(4x - 4) =$

4) $(2x - 5)(-6) =$

5) $(-3)(x + 6) =$

6) $(4 + 3x)2 =$

7) $(-5)(8 - 3x) =$

8) $-(-5 - 7x) =$

9) $(-6x + 3)(-3) =$

10) $(-4)(x - 7) =$

11) $-(5 - 3x) =$

12) $3(9 + 4x) =$

13) $6(4 + 3x) =$

14) $(-5x + 3)2 =$

15) $(5 - 8x)(-3) =$

16) $(-12)(3x + 3) =$

17) $(5 - 3x)6 =$

18) $4(2 + 6x) =$

19) $8(7x - 3) =$

20) $(-2x + 3)4 =$

21) $(7 - 5x)(-9) =$

22) $(-10)(x - 8) =$

23) $(11 - 4x)3 =$

24) $(-6)(10x - 4) =$

25) $(3 - 9x)(-7) =$

26) $(-9)(x + 9) =$

27) $(-3 + 5x)(-7) =$

28) $(-5)(8 - 10x) =$

29) $12(4x - 8) =$

30) $(-10x + 13)(-3) =$

31) $(-8)(3x - 2) + 4(x + 5) =$

32) $(-8)(x + 4) - (6 + 5x) =$

‖

Evaluating One Variable Expressions

✎ **Evaluate each expression using the value given.**

1) $8 - x, x = 5$

2) $x - 9, x = 5$

3) $5x + 4, x = 3$

4) $x - 13, x = -4$

5) $12 - x, x = 4$

6) $x + 2, x = 6$

7) $4x + 8, x = 3$

8) $x + (-7), x = -8$

9) $4x + 5, x = 2$

10) $3x + 9, x = -2$

11) $15 + 3x - 7, x = 2$

12) $17 - 3x, x = 3$

13) $8x - 9, x = 4$

14) $5x + 4, x = -3$

15) $10x + 5, x = 3$

16) $14 - 4x, x = -6$

17) $3(5x + 3), x = 9$

18) $4(-3x - 6), x = 3$

19) $7x - 2x + 12, x = 4$

20) $(5x + 6) \div 2, x = 8$

21) $(x + 18) \div 10, x = 12$

22) $5x - 12 + 3x, x = -3$

23) $(6 - 4x)(-3), x = -4$

24) $9x^2 + 3x - 6, x = 2$

25) $x^2 - 10x, x = -5$

26) $3x(7 - 2x), x = 2$

27) $12x + 6 - 2x^2, x = -4$

28) $(-3)(4x - 8 + 3x), x = 3$

29) $(-6) + \frac{x}{4} + 3x, x = 16$

30) $(-6) + \frac{x}{5}, x = 35$

31) $\left(-\frac{45}{x}\right) - 7 + 2x, x = 9$

32) $\left(-\frac{21}{x}\right) - 12 + 4x, x = 7$

Evaluating Two Variables Expressions

✎ **Evaluate each expression using the values given.**

1) $2x - 4y$,

   $x = 4, y = 1$

2) $3x + 5y$,

   $x = -2, y = 2$

3) $-7a + 4b$,

   $a = 2, b = 4$

4) $3x + 5 - y$,

   $x = 5, y = 6$

5) $3z + 12 - 2k$,

   $z = 5, k = 6$

6) $6(-x - 3y)$,

   $x = 5, y = -2$

7) $5a + 3b$,

   $a = 3, b = 4$

8) $7x \div 3y$,

   $x = 3, y = 7$

9) $2x + 15 + 5y$,

   $x = -3, y = 1$

10) $5a - (18 - b)$,

    $a = 2, b = 8$

11) $2z + 20 + 5k$,

    $z = -6, k = 5$

12) $xy + 10 + 4x$,

    $x = 3, y = 5$

13) $2x + 4y - 8 + 5$,

    $x = 5, y = 2$

14) $\left(-\frac{24}{x}\right) + 3 + 2y$,

    $x = 4, y = 6$

15) $(-3)(-3a - 3b)$,

    $a = 4, b = 5$

16) $12 + 4x - 7 - y$,

    $x = 3, y = 5$

17) $11x + 5 - 8y + 6$,

    $x = 5, y = 2$

18) $10 + 2(-4x - 5y)$,

    $x = 5, y = 4$

19) $5x + 13 + 6y$,

    $x = 5, y = 6$

20) $10a - (7a + 3b) - 11$,

    $a = 3, b = 8$

Combining like Terms

✏️ **Simplify each expression.**

1) $11x + 3x + 6 =$

2) $8(2x - 6) =$

3) $18x - 7x + 11 =$

4) $(-4)(6x - 7) =$

5) $22x - 10x - 5 =$

6) $32x - 13 + 8x =$

7) $15 - (8x - 11) =$

8) $-24x + 17 - 11x =$

9) $12x - 8 - 6x + 9 =$

10) $21x + 5 - 36 + 12x =$

11) $28x + 3x - 11 =$

12) $(-3x + 4)5 =$

13) $2 + 4x + 9x - 8 =$

14) $6(2x - 5x) - 4 =$

15) $4(5x + 11) + 3x =$

16) $x - 14 - 11x =$

17) $5(10 + 9x) - 8x =$

18) $42x + 17 - 23x =$

19) $(-7x) + 19 + 20x =$

20) $(-7x) - 33 + 29x =$

21) $4(5x + 3) - 19x =$

22) $5(6 - 2x) - 15x =$

23) $-24x + (11 - 18x) =$

24) $(-9) - (6)(7x + 3) =$

25) $(-1)(8x - 10) - 21x =$

26) $-36x + 14 + 27x - 5x =$

27) $3(-13x + 6) - 17x =$

28) $-5x - 42 + 32x =$

29) $37x - 19x + 15 - 9x =$

30) $3(5x + 7x) - 31 =$

31) $14 - 6x - 15 - 9x =$

32) $-2(-5x - 7x) + 27x =$

Answers of Worksheets

Simplifying Variable Expressions

1) $3x + 15$	12) $-11x^2 + 5x$	23) $4x^2 + 25x - 19$
2) $-28x + 20$	13) $-6x + 12$	24) $6x^2 - 66x + 25$
3) $5x + 5$	14) $x - 4$	25) $3x^2 + 19x - 5$
4) $-8x^2 - 4$	15) $22x - 5$	26) $-7x^2 - 20x$
5) $13x^2 + 10$	16) $14x - 22$	27) $-12x^2 + 10x - 7$
6) $18x^2 + 7x$	17) $27x - 16$	28) $-6x^2 - 21x + 13$
7) $-9x^2 + 4x$	18) $25x + 28$	29) $10x^2 + 7x + 17$
8) $4x^2 - 10x$	19) $-23x - 20$	30) $25x^2 + 25x$
9) $-22x + 21$	20) $-3x^2 - 14x$	31) $-16x^2 - 23x + 29$
10) $68x - 12$	21) $-20x^2 + 27x$	32) $-9x^2 + 3x + 30$
11) $-18x - 71$	22) $-20x^2 - 15x$	

Simplifying Polynomial Expressions

1) $2x^3 + 5x^2 - 11x$	9) $-3x^4 + 5x^3 + 2x^2 - 5x$
2) $2x^5 + 2x^3 - 11x^2$	10) $11x^4 - 3x^2 + 2x$
3) $21x^4 + x^2$	11) $-13x^4 + 3x^2$
4) $12x^3 + 8x^2 + 13x$	12) $19x^4 - 29x^3 + 2x^2$
5) $-26x^3 + 15x^2 - 22$	13) $3x^4 - 32x^3 + 10x^2$
6) $-5x^4 - 6x^3 - 3x$	14) $17x^3 - 13x^2 - 6x$
7) $-26x^3 - 10x^2 + 25x$	15) $-11x^5 - 29x^4 + 5x^2$
8) $-5x^3 - 11x^2 - 10x$	16) $10x^3 - 5x^2 + 13x$

Translate Phrases into an Algebraic Statement

1) $9x$	5) $y + 40$	9) $57 - y$	12) $x - 6 = 19$
2) $y - 11$	6) 6^2	10) $\frac{9}{x}$	13) $10a - b^2$
3) $\frac{19}{x}$	7) x^5	11) $\frac{x^2}{25}$	14) $41 - ab$
4) $38 - y$	8) $6 + x$		

The Distributive Property

1) $8x + 4$	4) $-12x + 30$	7) $15x - 40$	10) $-4x + 28$
2) $14x + 8$	5) $-3x - 18$	8) $7x + 5$	11) $3x - 5$
3) $12x - 12$	6) $6x + 8$	9) $18x - 9$	12) $12x + 27$

13) $18x + 24$	18) $24x + 8$	23) $-12x + 33$	28) $50x - 40$
14) $-10x + 6$	19) $56x - 24$	24) $-60x + 24$	29) $48x - 96$
15) $24x - 15$	20) $-8x + 12$	25) $63x - 21$	30) $30x - 39$
16) $-36x - 36$	21) $45x - 63$	26) $-9x - 81$	31) $-20x + 36$
17) $-18x + 30$	22) $-10x + 80$	27) $-35x + 21$	32) $-13x - 38$

Evaluating One Variables

1) 3	9) 13	17) 144	25) 75
2) -4	10) 3	18) -60	26) 18
3) 19	11) 14	19) 32	27) -74
4) -17	12) 8	20) 23	28) -39
5) 8	13) 23	21) 3	29) 46
6) 8	14) -11	22) -36	30) 1
7) 20	15) 35	23) -66	31) 6
8) -15	16) 38	24) 36	32) 13

Evaluating Two Variables

1) 4	6) 6	11) 33	16) 12
2) 4	7) 27	12) 37	17) 50
3) 2	8) 1	13) 15	18) -70
4) 14	9) 14	14) 9	19) 74
5) 15	10) 0	15) 81	20) -26

Combining like Terms

1) $14x + 6$	9) $6x + 1$	17) $37x + 50$	25) $-29x + 10$
2) $16x - 48$	10) $33x - 31$	18) $19x + 17$	26) $-14x + 14$
3) $11x + 11$	11) $31x - 11$	19) $13x + 19$	27) $-56x + 18$
4) $-24x + 28$	12) $-15x + 20$	20) $22x - 33$	28) $27x - 42$
5) $12x - 5$	13) $13x - 6$	21) $x + 12$	29) $9x + 15$
6) $40x - 13$	14) $-18x - 4$	22) $-25x + 30$	30) $36x - 31$
7) $-8x + 26$	15) $23x + 44$	23) $-42x + 11$	31) $-15x - 1$
8) $-35x + 17$	16) $-10x - 14$	24) $-42x - 27$	32) $51x$

Chapter 5 :

Exponents

Topics that you will practice in this chapter:

- ✓ Multiplication Property of Exponents
- ✓ Zero and Negative Exponents
- ✓ Division Property of Exponents
- ✓ Powers of Products and Quotients
- ✓ Negative Exponents and Negative Bases
- ✓ Scientific Notation

Love is anterior to life, posterior to death, initial of creation, and the exponent of breath.

Emily Dickinson

Multiplication Property of Exponents

✎ **Simplify and write the answer in exponential form.**

1) $4 \times 4^5 =$

2) $8^4 \times 8 =$

3) $7^3 \times 7^3 =$

4) $9^2 \times 9^2 =$

5) $2^2 \times 2^4 \times 2 =$

6) $5 \times 5^3 \times 5^3 =$

7) $4^3 \times 4^2 \times 4 \times 4 =$

8) $5x \times x =$

9) $x^3 \times x^3 =$

10) $x^7 \times x^2 =$

11) $x^4 \times x^3 \times x^2 =$

12) $10x \times 3x =$

13) $4x^3 \times 4x^3 =$

14) $7x^3 \times x =$

15) $3x^2 \times 4x^2 \times x^2 =$

16) $5x^4 \times x^4 =$

17) $2x^8 \times 2x =$

18) $6x \times x^5 =$

19) $4x^2 \times 6x^6 =$

20) $5yx^3 \times 4x =$

21) $7x^3 \times y^5x^7 =$

22) $y^2x^3 \times y^5x^4 =$

23) $3x^5 \times 4x^3y^4 =$

24) $4x^4 \times 9x^2y^5 =$

25) $5x^3y^4 \times 6x^8y^2 =$

26) $8x^3y^6 \times 4xy^3 =$

27) $2xy^5 \times 6x^3y^3 =$

28) $4x^5y^2 \times 4x^2y^8 =$

29) $7x \times 3y^8x^2 \times y^5 =$

30) $x^3 \times 2y^3x^4 \times 2y =$

31) $3yx^4 \times 3y^4x \times 3xy^3 =$

32) $6y^3 \times 2y^2x^4 \times 10yx^5 =$

Zero and Negative Exponents

✎ **Evaluate the following expressions.**

1) $1^{-5} =$

2) $4^{-1} =$

3) $0^{10} =$

4) $1^{15} =$

5) $5^{-2} =$

6) $3^{-3} =$

7) $9^{-1} =$

8) $10^{-2} =$

9) $12^{-2} =$

10) $2^{-5} =$

11) $3^{-4} =$

12) $2^{-4} =$

13) $6^{-3} =$

14) $10^{-3} =$

15) $30^{-1=}$

16) $15^{-2} =$

17) $4^{-3} =$

18) $2^{-7} =$

19) $5^{-3} =$

20) $4^{-4} =$

21) $3^{-5} =$

22) $10^{-4} =$

23) $2^{-10} =$

24) $8^{-3} =$

25) $20^{-2} =$

26) $14^{-2} =$

27) $9^{-3} =$

28) $100^{-2} =$

29) $5^{-4} =$

30) $4^{-6} =$

31) $\left(\frac{1}{4}\right)^{-3}$

32) $\left(\frac{1}{6}\right)^{-2} =$

33) $\left(\frac{1}{7}\right)^{-2} =$

34) $\left(\frac{2}{3}\right)^{-3} =$

35) $\left(\frac{1}{13}\right)^{-2} =$

36) $\left(\frac{7}{12}\right)^{-2} =$

37) $\left(\frac{1}{6}\right)^{-3} =$

38) $\left(\frac{1}{300}\right)^{-2} =$

39) $\left(\frac{2}{9}\right)^{-2} =$

40) $\left(\frac{7}{5}\right)^{-1} =$

41) $\left(\frac{13}{23}\right)^{0} =$

42) $\left(\frac{1}{4}\right)^{-5} =$

Division Property of Exponents

✏️ **Simplify.**

1) $\dfrac{5^6}{5^7} =$

2) $\dfrac{8^8}{8^6} =$

3) $\dfrac{4^5}{4} =$

4) $\dfrac{3}{3^5} =$

5) $\dfrac{x}{x^6} =$

6) $\dfrac{3 \times 3^2}{3^2 \times 3^5} =$

7) $\dfrac{9^4}{9^2} =$

8) $\dfrac{10 \times 10^9}{10^2 \times 10^7} =$

9) $\dfrac{7^5 \times 7^7}{7^4 \times 7^8} =$

10) $\dfrac{15x}{30x^6} =$

11) $\dfrac{3x^9}{4x^4} =$

12) $\dfrac{15x^8}{10x^9} =$

13) $\dfrac{42x^5}{6y^9} =$

14) $\dfrac{36y^8}{4x^4y^5} =$

15) $\dfrac{2x^7}{9x} =$

16) $\dfrac{49x^8y^6}{7x^9} =$

17) $\dfrac{48x^2}{24x^6y^{12}} =$

18) $\dfrac{30yx^5}{6yx^7} =$

19) $\dfrac{19x^7y}{38x^{12}y^4} =$

20) $\dfrac{9x^8}{63x^8} =$

21) $\dfrac{9x^{-9}}{4x^{-3}} =$

Powers of Products and Quotients

✎ **Simplify.**

1) $(4^3)^2 =$

2) $(2^3)^4 =$

3) $(2 \times 2^3)^2 =$

4) $(5 \times 5^5)^6 =$

5) $(19^4 \times 19^2)^3 =$

6) $(2^3 \times 2^4)^4 =$

7) $(5 \times 5^2)^2 =$

8) $(4^4)^4 =$

9) $(8x^5)^2 =$

10) $(3x^2y^4)^4 =$

11) $(7x^5y^2)^2 =$

12) $(5x^4y^4)^3 =$

13) $(2x^3y^3)^5 =$

14) $(10x^3y^4)^3 =$

15) $(13y^3y)^2 =$

16) $(5x^6x^4)^2 =$

17) $(6x^7y^6)^3 =$

18) $(12x^5x^7)^2 =$

19) $(2x^4 \times 2x)^4 =$

20) $(2x^4y^3)^5 =$

21) $(15x^7y^2)^2 =$

22) $(8x^3y^5)^3 =$

23) $(3x \times 2y^2)^4 =$

24) $\left(\frac{4x}{x^5}\right)^2 =$

25) $\left(\frac{x^4y^5}{x^3y^5}\right)^9 =$

26) $\left(\frac{36xy}{6x^5}\right)^3 =$

27) $\left(\frac{x^7}{x^8y^2}\right)^6 =$

28) $\left(\frac{xy^4}{x^3y^6}\right)^{-3} =$

29) $\left(\frac{5xy^8}{x^3}\right)^2 =$

30) $\left(\frac{xy^6}{2xy^3}\right)^{-4} =$

Negative Exponents and Negative Bases

✎ **Simplify.**

1) $-9^{-1} =$

2) $-9^{-2} =$

3) $-2^{-5} =$

4) $-x^{-7} =$

5) $11x^{-1} =$

6) $-8x^{-3} =$

7) $-12x^{-5} =$

8) $-9x^{-8}y^{-6} =$

9) $32x^{-5}y^{-1} =$

10) $10a^{-9}b^{-3} =$

11) $-17x^{4}y^{-6} =$

12) $-\dfrac{25}{x^{-5}} =$

13) $-\dfrac{13x}{a^{-7}} =$

14) $\left(-\dfrac{1}{3}\right)^{-4} =$

15) $\left(-\dfrac{3}{4}\right)^{-2} =$

16) $-\dfrac{14}{a^{-6}b^{-3}} =$

17) $-\dfrac{7x}{x^{-8}} =$

18) $-\dfrac{a^{-9}}{b^{-5}} =$

19) $-\dfrac{11}{x^{-5}} =$

20) $\dfrac{8b}{-16c^{-6}} =$

21) $\dfrac{12ab}{a^{-4}b^{-3}} =$

22) $-\dfrac{8n^{-4}}{32p^{-7}} =$

23) $\dfrac{16ab^{-6}}{-6c^{-5}} =$

24) $\left(\dfrac{10a}{5c}\right)^{-4} =$

25) $\left(-\dfrac{12x}{4yz}\right)^{-3} =$

26) $\dfrac{8ab^{-7}}{-5c^{-3}} =$

27) $\left(-\dfrac{x^{4}}{x^{5}}\right)^{-5} =$

28) $\left(-\dfrac{x^{-2}}{7x^{3}}\right)^{-2} =$

29) $\left(-\dfrac{x^{-4}}{x^{2}}\right)^{-6} =$

Scientific Notation

🖊 **Write each number in scientific notation.**

1) 0.223 =

2) 0.09 =

3) 4.5 =

4) 900 =

5) 2,000 =

6) 0.006 =

7) 33 =

8) 9,400 =

9) 1,470 =

10) 52,000 =

11) 8,000,000 =

12) 0.00009 =

13) 2,158,000 =

14) 0.0039 =

15) 0.000075 =

16) 4,300,000 =

17) 130,000 =

18) 4,000,000,000 =

19) 0.00009 =

20) 0.0039 =

🖊 **Write each number in standard notation.**

21) 4×10^{-1} =

22) 1.2×10^{-3} =

23) 2.7×10^{5} =

24) 6×10^{-4} =

25) 3.6×10^{-3} =

26) 5.5×10^{5} =

27) 3.2×10^{4} =

28) 3.88×10^{6} =

29) 7×10^{-6} =

30) 4.2×10^{-7} =

Answers of Worksheets

Multiplication Property of Exponents

1) 4^6
2) 8^5
3) 7^6
4) 9^4
5) 2^7
6) 5^7
7) 4^7
8) $5x^2$

9) x^6
10) x^9
11) x^9
12) $30x^2$
13) $16x^6$
14) $7x^4$
15) $12x^6$
16) $5x^8$

17) $4x^9$
18) $6x^6$
19) $24x^8$
20) $20x^4y$
21) $7x^{10}y^5$
22) x^7y^7
23) $12x^8y^4$
24) $36x^6y^5$

25) $30x^{11}y^6$
26) $32x^4y^9$
27) $12x^4y^8$
28) $16x^7y^{10}$
29) $21x^3y^{13}$
30) $4x^7y^4$
31) $27x^6y^8$
32) $120x^9y^6$

Zero and Negative Exponents

1) 1
2) $\frac{1}{4}$
3) 0
4) 1
5) $\frac{1}{25}$
6) $\frac{1}{27}$
7) $\frac{1}{9}$
8) $\frac{1}{100}$
9) $\frac{1}{144}$
10) $\frac{1}{32}$

11) $\frac{1}{81}$
12) $\frac{1}{16}$
13) $\frac{1}{216}$
14) $\frac{1}{1,000}$
15) $\frac{1}{30}$
16) $\frac{1}{225}$
17) $\frac{1}{64}$
18) $\frac{1}{128}$
19) $\frac{1}{125}$
20) $\frac{1}{256}$

21) $\frac{1}{243}$
22) $\frac{1}{10,000}$
23) $\frac{1}{1,024}$
24) $\frac{1}{512}$
25) $\frac{1}{400}$
26) $\frac{1}{196}$
27) $\frac{1}{729}$
28) $\frac{1}{10,000}$
29) $\frac{1}{625}$
30) $\frac{1}{4,096}$

31) 64
32) 36
33) 49
34) $\frac{27}{8}$
35) 169
36) $\frac{144}{49}$
37) 216
38) $90,000$
39) $\frac{81}{4}$
40) $\frac{5}{7}$
41) 1
42) $1,024$

Division Property of Exponents

1) $\frac{1}{5}$
2) 8^2
3) 4^4
4) $\frac{1}{3^4}$

5) $\frac{1}{x^5}$
6) $\frac{1}{3^4}$
7) 9^2
8) 10
9) 1

10) $\frac{1}{2x^5}$
11) $\frac{3x^5}{4}$
12) $\frac{3}{2x}$
13) $\frac{7x^5}{y^9}$

14) $\frac{9y^3}{x^4}$
15) $\frac{2x^6}{9}$
16) $\frac{7y^6}{x}$
17) $\frac{2}{x^4y^{12}}$

18) $\frac{5}{x^2}$ 19) $\frac{1}{2x^5y^3}$ 20) $\frac{1}{7}$ 21) $\frac{9}{4x^6}$

Powers of Products and Quotients

1) 4^6

2) 2^{12}

3) 2^8

4) 5^{36}

5) 19^{18}

6) 2^{28}

7) 5^6

8) 4^{16}

9) $64x^{10}$

10) $81x^8y^{16}$

11) $49x^{10}y^4$

12) $125x^{12}y^{12}$

13) $32x^{15}y^{15}$

14) $1,000x^9y^{12}$

15) $169y^8$

16) $25x^{20}$

17) $216x^{21}y^{18}$

18) $144x^{24}$

19) $256x^{20}$

20) $32x^{20}y^{15}$

21) $225x^{14}y^4$

22) $512x^9y^{15}$

23) $1,296x^4y^8$

24) $\frac{16}{x^8}$

25) x^9

26) $\frac{216y^3}{x^{12}}$

27) $\frac{1}{x^6y^{12}}$

28) x^6y^6

29) $\frac{25y^{16}}{x^4}$

30) $\frac{16}{y^{12}}$

Negative Exponents and Negative Bases

1) $-\frac{1}{9}$

2) $-\frac{1}{81}$

3) $-\frac{1}{32}$

4) $-\frac{1}{x^7}$

5) $\frac{11}{x}$

6) $-\frac{8}{x^3}$

7) $-\frac{12}{x^5}$

8) $-\frac{9}{x^8y^6}$

9) $\frac{32}{x^5y}$

10) $\frac{10}{a^9b^3}$

11) $-\frac{17x^4}{y^6}$

12) $-25x^5$

13) $-13xa^7$

14) 81

15) $\frac{16}{9}$

16) $-14a^6b^3$

17) $-7x^9$

18) $-\frac{b^5}{a^9}$

19) $-11x^5$

20) $-\frac{bc^6}{2}$

21) $12a^5b^4$

22) $-\frac{p^7}{4n^4}$

23) $-\frac{8ac^5}{3b^6}$

24) $\frac{c^4}{16a^4}$

25) $\frac{y^3z^3}{27x^3}$

26) $-\frac{8ac^3}{5b^7}$

27) $-x^5$

28) $49x^{10}$

29) x^{36}

Scientific Notation

1) 2.23×10^{-1}

2) 9×10^{-2}

3) 4.5×10^0

4) 9×10^2

5) 2×10^3

6) 6×10^{-3}

7) 3.3×10^1

8) 9.4×10^3

9) 1.47×10^3

10) 5.2×10^4

11) 8×10^6

12) 9×10^{-5}

13) 2.158×10^6

14) 3.9×10^{-3}

15) 7.5×10^{-5}

16) 4.3×10^6

17) 1.3×10^5

18) 4×10^9

19) 9×10^{-5}

20) 3.9×10^{-3}

21) 0.4

22) 0.0012

23) 270,000

24) 0.0006

25) 0.0036

26) 550,000

27) 32,000

28) 3,880,000

29) 0.000007

30) 0.00000042

Chapter 6 :

Radicals Expressions

Topics that you'll practice in this chapter:

✓ Square Roots

✓ Simplifying Radical Expressions

✓ Adding and Subtracting Radical Expressions

✓ Simplifying Radical Expressions Involving Fractions

✓ Multiplying Radical Expressions

"To be radical is to grasp things by the root."

— Karl Marx,

Square Roots

✎ **Find the value each square root.**

1) $\sqrt{16} =$ ___

2) $\sqrt{25} =$ ___

3) $\sqrt{1} =$ ___

4) $\sqrt{64} =$ ___

5) $\sqrt{0} =$ ___

6) $\sqrt{196} =$ ___

7) $\sqrt{4} =$ ___

8) $\sqrt{256} =$ ___

9) $\sqrt{36} =$ ___

10) $\sqrt{289} =$ ___

11) $\sqrt{169} =$ ___

12) $\sqrt{144} =$ ___

13) $\sqrt{100} =$ ___

14) $\sqrt{1,600} =$ ___

15) $\sqrt{2,500} =$ ___

16) $\sqrt{324} =$ ___

17) $\sqrt{529} =$ ___

18) $\sqrt{20} =$ ___

19) $\sqrt{625} =$ ___

20) $\sqrt{18} =$ ___

21) $\sqrt{50} =$ ___

22) $\sqrt{1,024} =$ ___

23) $\sqrt{160} =$ ___

24) $\sqrt{32} =$ ___

✎ **Evaluate.**

25) $\sqrt{4} \times \sqrt{25} =$ _____

26) $\sqrt{36} \times \sqrt{49} =$ _____

27) $\sqrt{6} \times \sqrt{6} =$ _____

28) $\sqrt{13} \times \sqrt{13} =$ _____

29) $2\sqrt{5} \times 3\sqrt{5} =$ _____

30) $\sqrt{12} \times \sqrt{3} =$ _____

31) $\sqrt{13} + \sqrt{13} =$ _____

32) $\sqrt{10} + 2\sqrt{10} =$ _____

33) $12\sqrt{7} - 10\sqrt{7} =$ _____

34) $4\sqrt{10} \times 2\sqrt{10} =$ _____

35) $5\sqrt{3} \times 8\sqrt{3} =$ _____

36) $6\sqrt{3} - \sqrt{12} =$ _____

Simplifying Radical Expressions

✎ **Simplify.**

1) $\sqrt{13x^2} =$

2) $\sqrt{75x^2} =$

3) $\sqrt[3]{27a} =$

4) $\sqrt{64x^5} =$

5) $\sqrt{216a} =$

6) $\sqrt[3]{63w^3} =$

7) $\sqrt{192x} =$

8) $\sqrt{125v} =$

9) $\sqrt[3]{128x^2} =$

10) $\sqrt{100x^9} =$

11) $\sqrt{16x^4} =$

12) $\sqrt[3]{500a^5} =$

13) $\sqrt{242} =$

14) $\sqrt{392p^3} =$

15) $\sqrt{8m^6} =$

16) $\sqrt{198x^3y^3} =$

17) $\sqrt{121x^5y^5} =$

18) $\sqrt{16a^6b^3} =$

19) $\sqrt{90x^5y^7} =$

20) $\sqrt[3]{64y^2x^6} =$

21) $10\sqrt{16x^4} =$

22) $6\sqrt{81x^2} =$

23) $\sqrt[3]{56x^2y^6} =$

24) $\sqrt[3]{1,000x^5y^7} =$

25) $8\sqrt{50a} =$

26) $\sqrt[4]{625x^8y} =$

27) $\sqrt{24x^4y^5r^3} =$

28) $5\sqrt{36x^4y^5z^8} =$

29) $3\sqrt[3]{343x^9y^7} =$

30) $5\sqrt{81a^5b^2c^9} =$

31) $\sqrt[4]{625x^8y^{16}} =$

Adding and Subtracting Radical Expressions

✎ **Simplify.**

1) $\sqrt{2} + \sqrt{8} =$

2) $3\sqrt{50} + 4\sqrt{2} =$

3) $2\sqrt{12} - 4\sqrt{3} =$

4) $5\sqrt{32} - 5\sqrt{2} =$

5) $3\sqrt{75} - 5\sqrt{3} =$

6) $-\sqrt{72} - 4\sqrt{2} =$

7) $-7\sqrt{16} - 4\sqrt{25} =$

8) $8\sqrt{24} + 2\sqrt{6} =$

9) $10\sqrt{49} - 7\sqrt{100} =$

10) $-7\sqrt{5} + 9\sqrt{45} =$

11) $-15\sqrt{12} + 14\sqrt{48} =$

12) $20\sqrt{4} - 2\sqrt{25} =$

13) $-2\sqrt{20} + 7\sqrt{5} =$

14) $8\sqrt{7} - 2\sqrt{63} =$

15) $5\sqrt{44} + 3\sqrt{11} =$

16) $3\sqrt{27} - 5\sqrt{48} =$

17) $\sqrt{144} - \sqrt{81} =$

18) $3\sqrt{20} - 6\sqrt{5} =$

19) $-2\sqrt{7} + 8\sqrt{28} =$

20) $3\sqrt{75} - 2\sqrt{3} =$

21) $5\sqrt{27} - 3\sqrt{3} =$

22) $-7\sqrt{30} + 6\sqrt{120} =$

23) $-7\sqrt{24} - 2\sqrt{6} =$

24) $-\sqrt{32x} + 4\sqrt{2x} =$

25) $\sqrt{7y^2} + y\sqrt{112} =$

26) $\sqrt{45mn^2} + 2n\sqrt{5m} =$

27) $-4\sqrt{12a} - 4\sqrt{3a} =$

28) $-5\sqrt{15ab} - 2\sqrt{60ab} =$

29) $\sqrt{45x^2y} + x\sqrt{20y} =$

30) $2\sqrt{7a} + 4\sqrt{63a} =$

Multiplying Radical Expressions

✎ **Simplify.**

1) $\sqrt{5} \times \sqrt{5} =$

2) $\sqrt{5} \times \sqrt{10} =$

3) $\sqrt{3} \times \sqrt{12} =$

4) $\sqrt{49} \times \sqrt{47} =$

5) $\sqrt{7} \times -2\sqrt{28} =$

6) $3\sqrt{15} \times \sqrt{5} =$

7) $4\sqrt{72} \times \sqrt{2} =$

8) $\sqrt{5} \times -\sqrt{49} =$

9) $\sqrt{55} \times \sqrt{11} =$

10) $7\sqrt{42} \times 2\sqrt{216} =$

11) $\sqrt{45}(5 + \sqrt{5}) =$

12) $\sqrt{13x^2} \times \sqrt{13x^3} =$

13) $-2\sqrt{27} \times \ \sqrt{3} =$

14) $2\sqrt{13x^4} \times \sqrt{13x^4} =$

15) $\sqrt{14x^3} \times \sqrt{7x^2} =$

16) $-8\sqrt{5x} \times \sqrt{7x^5} =$

17) $-2\sqrt{16x^5} \times 4\sqrt{8x^3} =$

18) $-4\sqrt{32}(8 + \sqrt{32}) =$

19) $\sqrt{32x}\,(10 - \sqrt{2x}) =$

20) $\sqrt{2x}(8\sqrt{x^5} + \sqrt{8}) =$

21) $\sqrt{20r}\,(5 + \sqrt{5}) =$

22) $-4\sqrt{7x} \times 3\sqrt{14x^5} =$

23) $-2\sqrt{12x} \times 3\sqrt{2x}$

24) $-\sqrt{7v^3}\,(-3\sqrt{42v}) =$

25) $(\sqrt{11} - 5)(\sqrt{11} + 5) =$

26) $(-3\sqrt{5} + 3)(\sqrt{5} - 4) =$

27) $(4 - 6\sqrt{3})(-6 + \sqrt{3}) =$

28) $(8 - 3\sqrt{5})(7 - \sqrt{5}) =$

29) $(-1 - \sqrt{3x})(\,4 + \sqrt{3x}) =$

30) $(-5 + 2\sqrt{7r})(-5 + \sqrt{7r}) =$

31) $(-\sqrt{7n} + 1)(-\sqrt{7} - 5) =$

32) $(-3 + \sqrt{3})(\,5 - 2\sqrt{3x}) =$

Simplifying Radical Expressions Involving Fractions

✎ **Simplify.**

1) $\dfrac{\sqrt{5}}{\sqrt{3}} =$

2) $\dfrac{\sqrt{18}}{\sqrt{45}} =$

3) $\dfrac{\sqrt{10}}{5\sqrt{2}} =$

4) $\dfrac{13}{\sqrt{3}} =$

5) $\dfrac{12\sqrt{5r}}{\sqrt{m^5}} =$

6) $\dfrac{11\sqrt{2}}{\sqrt{k}} =$

7) $\dfrac{6\sqrt{20x^3}}{\sqrt{16x}} =$

8) $\dfrac{\sqrt{14x^3y^4}}{\sqrt{7x^4y^3}} =$

9) $\dfrac{1}{1-\sqrt{5}} =$

10) $\dfrac{1-8\sqrt{a}}{\sqrt{11a}} =$

11) $\dfrac{\sqrt{a}}{\sqrt{a}+\sqrt{b}} =$

12) $\dfrac{1-\sqrt{5}}{2-\sqrt{6}} =$

13) $\dfrac{4+\sqrt{7}}{3-\sqrt{8}} =$

14) $\dfrac{5}{-3-3\sqrt{3}} =$

15) $\dfrac{7}{2-\sqrt{5}} =$

16) $\dfrac{\sqrt{7}-\sqrt{3}}{\sqrt{3}-\sqrt{7}} =$

17) $\dfrac{\sqrt{5}+\sqrt{7}}{\sqrt{7}-\sqrt{5}} =$

18) $\dfrac{2\sqrt{2}-\sqrt{3}}{3\sqrt{2}+\sqrt{5}} =$

19) $\dfrac{\sqrt{11}+5\sqrt{3}}{4-\sqrt{11}} =$

20) $\dfrac{\sqrt{5}+\sqrt{3}}{2-\sqrt{3}} =$

21) $\dfrac{\sqrt{32a^7b^4}}{\sqrt{2ab^3}} =$

22) $\dfrac{10\sqrt{21x^5}}{5\sqrt{x^3}} =$

Answers of Worksheets

Square Roots

1) 4	10) 17	19) 25	28) 13
2) 5	11) 13	20) $3\sqrt{2}$	29) 30
3) 1	12) 12	21) $5\sqrt{2}$	30) 6
4) 8	13) 10	22) 32	31) $2\sqrt{13}$
5) 0	14) 40	23) $4\sqrt{10}$	32) $3\sqrt{10}$
6) 14	15) 50	24) $4\sqrt{2}$	33) $2\sqrt{7}$
7) 2	16) 18	25) 10	34) 80
8) 16	17) 23	26) 42	35) 120
9) 6	18) $2\sqrt{5}$	27) 6	36) $4\sqrt{3}$

Simplifying radical expressions

1) $x\sqrt{13}$	12) $5a\sqrt[3]{4a^2}$	23) $2y^2\sqrt[3]{7x^2}$
2) $5x\sqrt{3}$	13) $11\sqrt{2}$	24) $10xy^2\sqrt[3]{x^2y}$
3) $3\sqrt[3]{a}$	14) $14p\sqrt{2p}$	25) $40\sqrt{2a}$
4) $8x^2\sqrt{x}$	15) $2m^3\sqrt{2}$	26) $5x^2\sqrt[4]{y}$
5) $6\sqrt{6a}$	16) $3x.y\sqrt{22xy}$	27) $2x^2y^2r\sqrt{6yr}$
6) $w\sqrt[3]{63}$	17) $11x^2y^2\sqrt{xy}$	28) $30x^2y^2z^4\sqrt{y}$
7) $8\sqrt{3x}$	18) $4a^3b\sqrt{b}$	29) $21x^3y^2\sqrt[3]{y}$
8) $5\sqrt{5v}$	19) $3x^2y^3\sqrt{10xy}$	30) $45a^2bc^4\sqrt{ac}$
9) $4\sqrt[3]{2x^2}$	20) $4x^2\sqrt[3]{y^2}$	31) $5x^2y^4$
10) $10x^4\sqrt{x}$	21) $40x^2$	
11) $4x^2$	22) $54x$	

Adding and subtracting radical expressions

1) $3\sqrt{2}$	6) $-10\sqrt{2}$	11) $26\sqrt{3}$
2) $19\sqrt{2}$	7) -48	12) 30
3) 0	8) $18\sqrt{6}$	13) $3\sqrt{5}$
4) $15\sqrt{2}$	9) 0	14) $2\sqrt{7}$
5) $10\sqrt{3}$	10) $20\sqrt{5}$	15) $13\sqrt{11}$

16) $-11\sqrt{3}$

17) 3

18) 0

19) $14\sqrt{7}$

20) $13\sqrt{3}$

21) $12\sqrt{3}$

22) $5\sqrt{30}$

23) $-16\sqrt{6}$

24) 0

25) $5y\sqrt{7}$

26) $5n\sqrt{5m}$

27) $-12\sqrt{3a}$

28) $-9\sqrt{15ab}$

29) $5x\sqrt{5y}$

30) $14\sqrt{7a}$

Multiplying radical expressions

1) 5

2) $5\sqrt{2}$

3) 6

4) $7\sqrt{47}$

5) -28

6) $15\sqrt{3}$

7) 48

8) $-5\sqrt{7}$

9) $11\sqrt{5}$

10) $504\sqrt{7}$

11) $15\sqrt{5} + 15$

12) $13x^2\sqrt{x}$

13) -18

14) $26x^4$

15) $7x^2\sqrt{2x}$

16) $-8x^3\sqrt{35}$

17) $-64x^4\sqrt{2}$

18) $-128\sqrt{2} - 128$

19) $40\sqrt{2x} - 8x$

20) $8x^3\sqrt{2} + 4\sqrt{x}$

21) $10\sqrt{5r} + 10\sqrt{r}$

22) $-84x^3\sqrt{2}$

23) $-12\sqrt{6}x$

24) $21v^2\sqrt{6}$

25) -14

26) $15\sqrt{5} - 27$

27) $40\sqrt{3} - 42$

28) $71 - 29\sqrt{5}$

29) $-3x - 5\sqrt{3x} - 4$

30) $14r - 15\sqrt{7r} + 25$

31) $7\sqrt{n} + 5\sqrt{7n} - \sqrt{7} - 5$

32) $-15 + 6\sqrt{3x} + 5\sqrt{3} - 6\sqrt{x}$

Simplifying radical expressions involving fractions

1) $\frac{\sqrt{15}}{3}$

2) $\frac{9\sqrt{10}}{45} = \frac{\sqrt{10}}{5}$

3) $\frac{\sqrt{20}}{10} = \frac{\sqrt{5}}{5}$

4) $\frac{13\sqrt{3}}{3}$

5) $\frac{12\sqrt{5mr}}{m^3}$

6) $\frac{11\sqrt{2k}}{k}$

7) $3x\sqrt{5}$

8) $\frac{\sqrt{2x}}{xy}$

9) $\frac{-1-\sqrt{5}}{4}$

10) $\frac{\sqrt{11a}-8a\sqrt{11}}{11a}$

11) $\frac{a-\sqrt{ab}}{a-b}$

12) $\frac{\sqrt{30}+2\sqrt{5}-\sqrt{6}-2}{2}$

13) $12 + 8\sqrt{2} + 3\sqrt{7} + 2\sqrt{14}$

14) $-\frac{5(\sqrt{3}-1)}{6}$

15) $-14 - 7\sqrt{5}$

16) -1

17) $6 + \sqrt{35}$

18) $\frac{12 - 2\sqrt{10} - 3\sqrt{6} + \sqrt{15}}{13}$

19) $\frac{4\sqrt{11}+11+20\sqrt{3}+5\sqrt{33}}{5}$

20) $2\sqrt{5} + 3 + \sqrt{15} + 2\sqrt{3}$

21) $4a^3\sqrt{b}$

22) $2x\sqrt{21}$

Chapter 7 :

Linear Functions

Topics that you'll practice in this chapter:

✓ Finding Slope,

✓ Graphing Linear Equations,

✓ Graphing Linear Inequalities,

✓ Writing Linear Equations,

✓ Graphing Horizontal and Vertical lines

✓ Finding Rate of Change,

✓ Find the x–intercept and y–intercept,

✓ Slope-Intercept Form Equations,

✓ Point-Slope Form Equations,

✓ Equation of parallel or perpendicular lines,

✓ Graphing Lines of Equations,

✓ Graphing Absolute Value Equations

Life is not linear; you have ups and downs. It's how you deal with the troughs that defines you.
Michael Lee-Chin

Finding Slope

🔖 **Find the slope of each line.**

1) $y = x + 8$

2) $y = -3x + 5$

3) $y = 2x + 12$

4) $y = -4x + 19$

5) $y = 11 + 6x$

6) $y = 7 - 5x$

7) $y = 8x + 19$

8) $y = -9x + 20$

9) $y = -7x + 4$

10) $y = 3x - 8$

11) $y = \frac{1}{3}x + 8$

12) $y = -\frac{4}{5}x + 9$

13) $-3x + 6y = 30$

14) $4x + 4y = 16$

15) $3y - x = 10$

16) $8y - x = 5$

🔖 **Find the slope of the line through each pair of points.**

17) $(2, 3), (7, 10)$

18) $(-3, 5), (2, 15)$

19) $(5, -3), (1, 9)$

20) $(-5, -5), (10, 25)$

21) $(22, 3), (7, 18)$

22) $(-16, 8), (-7, 26)$

23) $(25, 11), (29, 19)$

24) $(26, -19), (14, 17)$

25) $(22, -13), (20, -11)$

26) $(19, 7), (15, -3)$

27) $(5, 7), (11, 19)$

28) $(52, -62), (40, 70)$

Graphing Linear Equations

✎ **Sketch the graph of each line.**

1) $y = x - 2$ 2) $y = -3x + 2$ 3) $x + y = 0$

  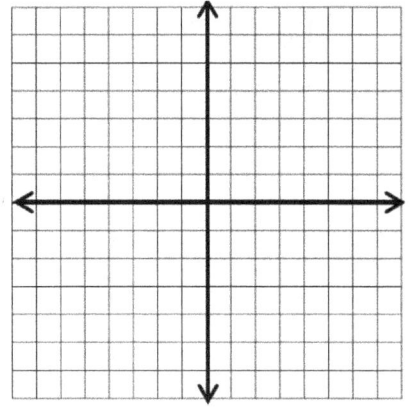

4) $x + y = -3$ 5) $2x + 3y = -4$ 6) $y - 3x + 6 = 0$

  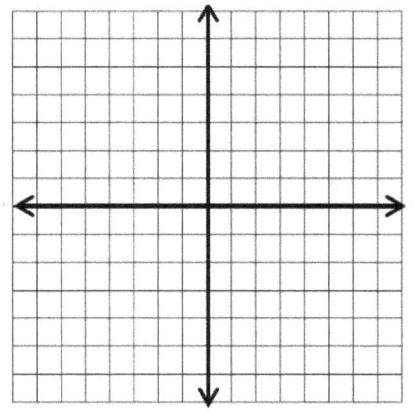

Graphing Linear Inequalities

✎ **Sketch the graph of each linear inequality.**

1) $y > 4x - 5$ 2) $y < 2x + 4$ 3) $y \leq -5x - 2$

  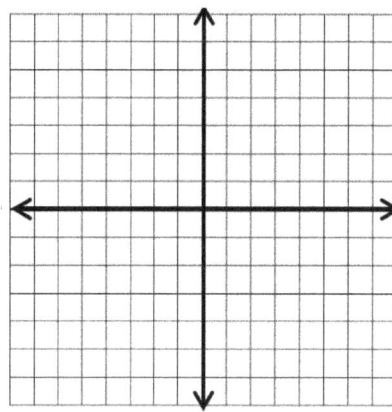

4) $4y \geq 12 + 4x$ 5) $-12y < 3x - 24$ 6) $5y \geq -15x + 10$

  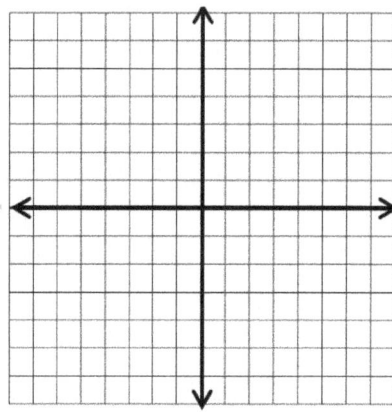

Writing Linear Equations

✍ **Write the equation of the line through the given points.**

1) Through: $(2, -5), (3, 9)$

2) Through: $(-6, 3), (3, 12)$

3) Through: $(10, 7), (5, 27)$

4) Through: $(15, 11), (3, -1)$

5) Through: $(24, 17), (12, -7)$

6) Through: $(8, 29), (4, -7)$

7) Through: $(20, -16), (12, 0)$

8) Through: $(-3, 10), (2, -5)$

9) Through: $(-6, 17), (4, -3)$

10) Through: $(-8, 22), (5, -4)$

11) Through: $(9, 27), (3, -3)$

12) Through: $(11, 32), (9, 4)$

13) Through: $(-3, 13), (-4, 0)$

14) Through: $(-5, 5), (5, 15)$

15) Through: $(18, -32), (11, 3)$

16) Through: $(-4, 25), (4, -15)$

✍ **Find the answer for each problem.**

17) What is the equation of a line with slope 6 and intercept 12?

_____

18) What is the equation of a line with slope -11 and intercept -4?

_____

19) What is the equation of a line with slope -3 and passes through point $(5, 2)$? _____

20) What is the equation of a line with slope -5 and passes through point $(-2, -1)$? _____

21) The slope of a line is -10 and it passes through point $(-3, 0)$. What is the equation of the line? _____

22) The slope of a line is 8 and it passes through point $(0, 7)$. What is the equation of the line? _____

TSI Subject Test Mathematics

✎ **Find the value of** b: **The line that passes through each pair of points has the given slope.**

1) $(5, -4), (2, b), m = 1$

2) $(b, -4), (-4, 1), m = -\frac{1}{3}$

3) $(-4, b), (4, 8), m = \frac{1}{2}$

4) $(0, 3), (b, 8), m = 1\frac{2}{3}$

✎ **Write the slope intercept form of the equation of each line.**

1)

2)

3)

4)

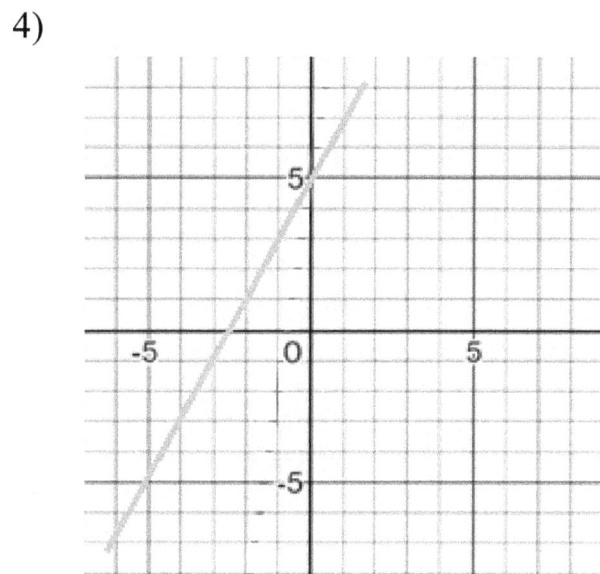

Graphing Horizontal and Vertical Lines

✎ **Sketch the graph of each line.**

1) $y = 3$

2) $y = -1$

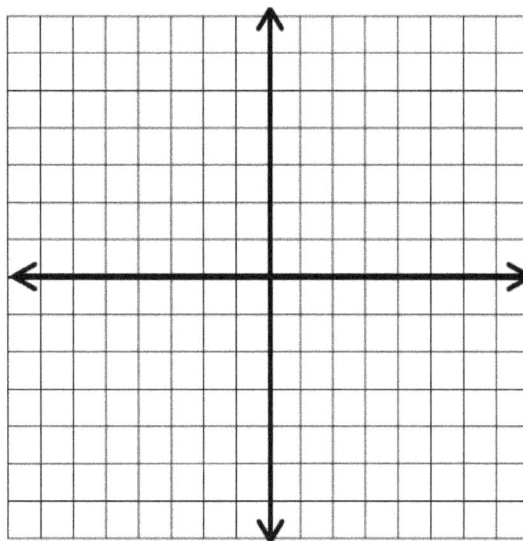

3) $x = 0$

4) $x = 3$

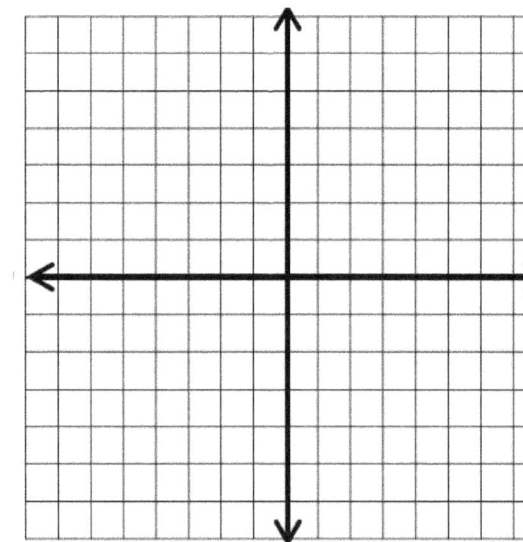

Rate of change

✍ **What is the average rate of change of the function?**

1) $f(x) = 3x^2 + 5$, from $x = 3$ to $x = 6$?

2) $f(x) = -2x^2 - 4$, from $x = 2$ to $x = 4$?

3) $f(x) = x^3 + 3$, from $x = 1$ to $x = 2$?

x and *y* intercepts

✍ **Find the x and y intercepts for the following equations.**

1) $5x + 3y = 15$

2) $y = x + 8$

3) $4x = y + 16$

4) $x + y = -2$

5) $4x - 3y = 7$

6) $7y - 5x + 10 = 0$

7) $\frac{3}{7}x + \frac{1}{4}y + \frac{2}{3} = 0$

8) $3x - 21 = 0$

9) $24 - 4y = 0$

10) $-2x - 6y + 42 = 12$

Slope–intercept Form

✎ **Write the slope–intercept form of the equation of each line.**

1) $-14x + y = 6$

2) $-2(7x + y) = 24$

3) $-8x - 16y = -48$

4) $5x + 14 = -3y$

5) $x - 3y = 12$

6) $18x - 12y = -6$

7) $28x - 14y = -56$

8) $7x - 4y + 25 = 0$

9) $-\frac{1}{3}y = -2x + 3$

10) $5 - y - 4x = 0$

11) $-y = -6x - 9$

12) $10x + 5y = -15$

13) $3(x + y + 2) = 0$

14) $y - 4 = x + 3$

15) $3(y + 3) = 2(x - 3)$

16) $\frac{3}{4}y + \frac{1}{4}x + \frac{5}{4} = 0$

Point–slope Form

✍ **Find the slope of the following lines. Name a point on each line.**

1) $y = 2(x + 3)$

2) $y + 4 = \dfrac{1}{3}(x - 1)$

3) $y + 3 = -1.5x$

4) $y - 3 = \dfrac{1}{2}(x - 2)$

5) $y + 2 = 0.4(x + 3)$

6) $y - 8 = -3x$

7) $y - 12 = -3(x - 8)$

8) $y + 14 = 0$

9) $y + 18 = 2(x + 5)$

10) $y - 17 = -8(x - 3)$

✍ **Write an equation in point–slope form for the line that passes through the given point with the slope provided.**

11) $(2, -3), m = 4$

12) $(-7, 4), m = \dfrac{1}{5}$

13) $(0, -6), m = -2$

14) $(-a, b), m = m$

15) $(-9, 1), m = 3$

16) $(3, 0), m = -5$

17) $(-4, 11), m = \dfrac{1}{3}$

18) $(0, 11), m = 0$

19) $\left(-\dfrac{1}{3}, 3\right), m = \dfrac{1}{5}$

20) $(0, 0), m = -3$

Equation of Parallel or Perpendicular Lines

✍ **Write an equation of the line that passes through the given point and is parallel to the given line.**

1) $(-1, -2), x + 3y = -11$

2) $(-4, 1), y = x - 5$

3) $(-2, 0), 2y = 5x - 3$

4) $(0, 0), -3y + 4x - 14 = 0$

5) $(1, 10), y + 15 = 0$

6) $(0, 7), -5x - y = -4$

7) $(-2, -1), y = \frac{4}{5}x + 3$

8) $(-2, 5), -8x + 5y = -18$

9) $(3, -2), y = -\frac{2}{5}x - 3$

10) $(-5, -5), 6x + 15y = -30$

✍ **Write an equation of the line that passes through the given point and is perpendicular to the given line.**

11) $(-2, -6), 3x + 4y = -8$

12) $(-\frac{1}{3}, \frac{3}{5}), 4x - 8y = -32$

13) $(2, -5), y = -5$

14) $(7, -2), x = 7$

15) $(0, -3), y = \frac{1}{2}x + 6$

16) $(\frac{3}{5}, \frac{2}{5}), y = -6x - 24$

17) $(-10, 0), y = \frac{5}{3}x - 15$

18) $(3, -5), y = x + 12$

19) $(-3, -1), y = \frac{7}{3}x - 4$

20) $(0, 0), y - 8x + 6 = 0$

Graphing Absolute Value Equations

✎ **Graph each equation.**

1) $y = |x + 4|$ 2) $y = |x + 1|$ 3) $y = -|x| - 1$

  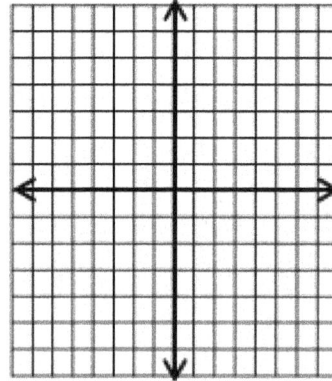

4) $y = |x - 2|$ 5) $y = -|x - 2|$ 6) $y = -2|2x + 2| + 4$

Answers of Worksheets

Finding Slope

1) 1

2) −3

3) 2

4) −4

5) 6

6) −5

7) 8

8) −9

9) −7

10) 3

11) $\frac{1}{3}$

12) $-\frac{4}{5}$

13) $\frac{1}{2}$

14) −1

15) $\frac{1}{3}$

16) $\frac{1}{8}$

17) $\frac{7}{5}$

18) 2

19) −3

20) 2

21) −1

22) 2

23) 2

24) −3

25) −1

26) $\frac{5}{2}$

27) 2

28) −11

Graphing Linear Equations

1) $y = x - 2$

2) $y = -3x + 2$

3) $x + y = 0$

4) $x + y = -3$

5) $2x + 3y = -4$

6) $y - 3x + 6 = 0$

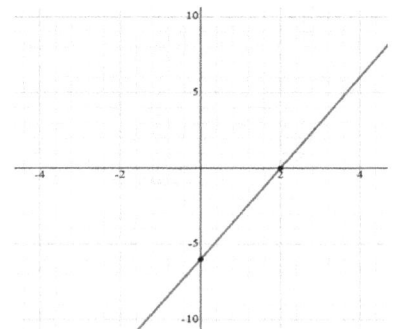

Graphing Linear Inequalities

1) $y > 4x - 5$

2) $y < 2x + 4$

3) $y \leq -5x - 2$

4) $4y \geq 12 + 4x$

5) $-12y < 3x - 24$

6) $5y \geq -15x + 10$

Writing Linear Equations

1) $y = 14x - 33$

2) $y = x + 9$

3) $y = -4x + 47$

4) $y = x - 4$

5) $y = 2x - 31$

6) $y = 9x - 43$

7) $y = -2x + 24$

8) $y = -3x + 1$

9) $y = -2x + 5$

10) $y = -2x + 6$

11) $y = 5x - 18$

12) $y = 14x - 122$

13) $y = 13x + 52$

14) $y = x + 10$

15) $y = -5x + 58$

16) $y = -5x + 5$

17) $y = 6x + 12$

18) $y = -11x - 4$

19) $y = -3x + 17$

20) $y = -5x - 11$

21) $y = -10x - 30$

22) $y = 8x + 7$

Find the value of b

1) -7

2) 11

3) 4

4) 3

Write an equation from a graph

1) $y = 2x + 4$

2) $y = 3x + 5$

3) $y = 5x$

4) $y = 2x + 5$

Graphing horizontal and vertical lines

1) $y = 3$

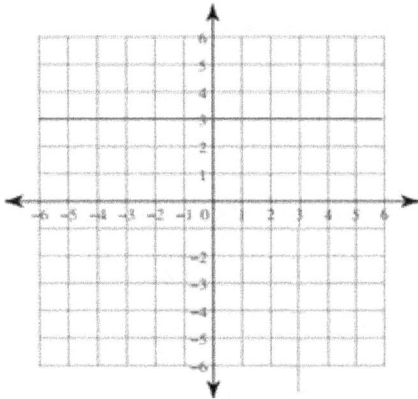

2) $y = -1$(it is on x axes)

3) $x = 0$

4) $x = 3$

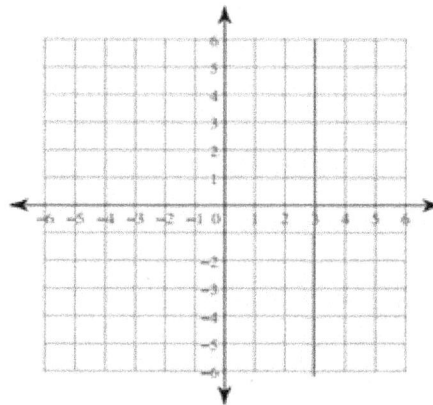

Rate of change

1) 27   2) −12   3) 7

x–intercept and y–intercept

1) y − intercept = 5   x − intercept = 3

2) y − intercept = 8   x − intercept = −8

3) y − intercept = −16   x − intercept = 4

4) y − intercept = −2   x − intercept = −2

5) y − intercept = $-\frac{7}{3}$ x − intercept = $\frac{7}{4}$

6) y − intercept = $-\frac{10}{7}$ x − intercept = 2

7) y − intercept = $-\frac{8}{3}$ x − intercept = $-\frac{2}{7}$

8) y − intercept = undefind   x − intercept = 7

9) $y -$ intercept $= 6$ $x -$ intercept $=$ undefind

10) $y -$ intercept $= 5$ $x -$ intercept $= 15$

Slope–intercept form

1) $y = 14x + 6$

2) $y = -7x - 12$

3) $y = -\frac{1}{2}x + 3$

4) $y = -\frac{5}{3}x - \frac{14}{3}$

5) $y = \frac{x}{3} - 4$

6) $y = \frac{3}{2}x + \frac{1}{2}$

7) $y = 2x + 4$

8) $y = \frac{7}{4}x + \frac{25}{4}$

9) $y = 6x - 9$

10) $y = -4x + 5$

11) $y = 6x + 9$

12) $y = -2x - 3$

13) $y = -x - 2$

14) $y = x + 7$

15) $y = \frac{2}{3}x - 5$

16) $y = -\frac{1}{3}x - \frac{5}{3}$

Point–slope form

1) $m = 2, (-3, 0)$

2) $m = \frac{1}{3}, (1, -4)$

3) $m = -\frac{3}{2}, (0, -3)$

4) $m = \frac{1}{2}, (2, 3)$

5) $m = \frac{4}{10}, (-3, -2)$

6) $m = -3, (0, 8)$

7) $m = -3, (8, 12)$

8) $m = 0, (0, -14)$

9) $m = 2, (-5, -18)$

10) $m = -8, (-3, 17)$

11) $y + 3 = 4(x - 2)$

12) $y - 4 = \frac{1}{5}(x + 7)$

13) $y + 6 = -2x$

14) $y - b = m(x + a)$

15) $y - 1 = 3(x + 9)$

16) $y = -5(x - 3)$

17) $y - 11 = \frac{1}{3}(x + 4)$

18) $y - 11 = 0$

19) $y - 3 = \frac{1}{5}\left(x + \frac{1}{3}\right)$

20) $y = -3x$

Equation of Parallel or Perpendicular Lines.

1) $y = -\frac{1}{3}x - 2\frac{1}{3}$

2) $y = x + 5$

3) $y = \frac{5}{2}x + 5$

4) $y = \frac{4}{3}x$

5) $y = 10$

6) $y = -5x + 7$

7) $y = \frac{4}{5}x + \frac{3}{5}$

8) $y = \frac{8}{5}x + \frac{41}{5}$

9) $y = -\frac{2}{5}x - \frac{4}{5}$

10) $y = -\frac{2}{5}x - 7$

11) $y = \frac{4}{3}x - \frac{10}{3}$

12) $y = -2x - \frac{1}{15}$

13) $x = 2$

14) $y = -2$

15) $y = -2x - 3$

16) $y = \frac{1}{6}x + \frac{3}{10}$

17) $y = -\frac{3}{5}x - 6$

18) $y = -x - 2$

19) $y = -\frac{3}{7}x - \frac{16}{7}$

20) $y = -\frac{1}{8}x$

Graphing Absolute Value Equations

1) y = |x + 4|

2) y = |x − 1|

3) y = − |x| − 1

4) y = |x − 2|

5) y = − |x − 2|

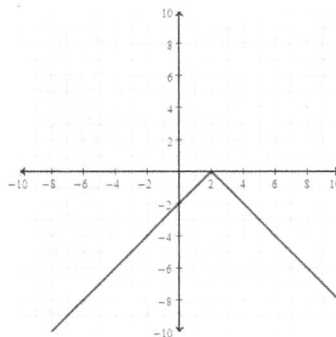

6) y = −2 |2x + 2| + 4

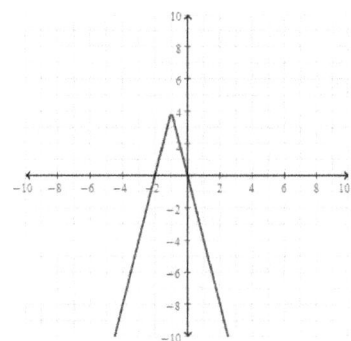

Chapter 8 :

Equations and Inequalities

Topics that you'll practice in this chapter:

- ✓ One–Step Equations
- ✓ Multi–Step Equations
- ✓ Graphing Single–Variable Inequalities
- ✓ One–Step Inequalities
- ✓ Multi-Step Inequalities
- ✓ Systems of Equations
- ✓ Systems of Equations Word Problems
- ✓ Finding Midpoint
- ✓ Finding Distance of Two Points

"Life is a math equation. In order to gain the most, you have to know how to convert negatives into positives." – Anonymous

One–Step Equations

✎ **Find the answer for each equation.**

1) $3x = 90, x =$ ___

2) $5x = 35, x =$ ___

3) $6x = 24, x =$ ___

4) $24x = 144, x =$ ___

5) $x + 15 = 20, x =$ ___

6) $x - 7 = 4, x =$ ___

7) $x - 9 = 2, x =$ ___

8) $x + 15 = 23, x =$ ___

9) $x - 4 = 13, x =$ ___

10) $12 = 16 + x, x =$ ___

11) $x - 10 = 2, x =$ ___

12) $5 - x = -11, x =$ ___

13) $28 = -6 + x, x =$ ___

14) $x - 20 = -35, x =$ ___

15) $x + 14 = -4, x =$ ___

16) $14 = 28 - x, x =$ ___

17) $7 + x = -7, x =$ ___

18) $x - 16 = 4, x =$ ___

19) $30 = x - 15, x =$ ___

20) $x - 5 = -18, x =$ ___

21) $x - 10 = 24, x =$ ___

22) $x - 20 = -25, x =$ ___

23) $x - 17 = 30, x =$ ___

24) $-70 = x - 28, x =$ ___

25) $x - 9 = 13, x =$ ___

26) $36 = 4x, x =$ ___

27) $x - 35 = 25, x =$ ___

28) $x - 25 = 10, x =$ ___

29) $70 - x = 16, x =$ ___

30) $x - 10 = 14, x =$ ___

31) $17 - x = -13, x =$ __

32) $x - 9 = -30, x =$ ___

Multi–Step Equations

✎ **Find the answer for each equation.**

1) $3x + 3 = 9$

2) $-x + 5 = 12$

3) $4x - 8 = 8$

4) $-(3 - x) = 5$

5) $4x - 8 = 16$

6) $12x - 15 = 9$

7) $2x - 18 = 2$

8) $4x + 8 = 16$

9) $24x + 27 = 75$

10) $-14(3 + x) = 14$

11) $-3(2 + x) = 6$

12) $12 = -(x - 7)$

13) $3(3 - x) = 30$

14) $-15 = -(3x + 6)$

15) $40(3 + x) = 40$

16) $5(x - 10) = 25$

17) $-18 = x + 8x$

18) $3x + 25 = -2x - 10$

19) $7(6 + 3x) = -63$

20) $18 - 3x = -4 - 5x$

21) $4 - 6x = 36 + 2x$

22) $15 + 15x = -5 + 5x$

23) $42 = (-6x) - 7 + 7$

24) $21 = 3x - 21 + 4x$

25) $-18 = -6x - 9 + 3x$

26) $5x - 15 = -29 + 6x$

27) $7x - 18 = 4x + 3$

28) $-7 - 4x = 5(4 - x)$

29) $x - 5 = -5(-3 - x)$

30) $13x - 68 = 15x - 102$

31) $-5x - 3 = -3(9 + 3x)$

32) $-2x - 15 = 6x + 17$

Graphing Single–Variable Inequalities

✍ **Draw a graph for each inequality.**

1) $x > -1$

2) $x \leq 2$

3) $x \geq 0$

4) $x < -3$

5) $x < \frac{1}{2}$

6) $x \leq -2$

7) $x \leq 3$

8) $x \geq -\frac{7}{2}$

One–Step Inequalities

✎ **Find the answer for each inequality and graph it.**

1) $x + 4 \geq 4$

2) $x - 5 \leq 2$

3) $5x > 35$

4) $9 + x \leq 11$

5) $x - 5 < -9$

6) $9x \geq 72$

7) $9x \leq 27$

8) $x + 19 > 16$

Multi-Step Inequalities

✍ **Calculate each inequality.**

1) $x - 3 \le 7$

2) $8 - x \le 8$

3) $3x - 9 \le 9$

4) $4x - 4 \ge 8$

5) $x - 7 \ge 1$

6) $5x - 15 \le 5$

7) $6x - 8 \le 4$

8) $-11 + 6x \le 12$

9) $4(x - 4) \le 16$

10) $3x - 10 \le 11$

11) $5x - 25 < 25$

12) $9x - 5 < 22$

13) $20 - 7x \ge -15$

14) $33 + 6x < 45$

15) $8 + 8x \ge 96$

16) $7 + 3x < 13$

17) $4x - 3 < 9$

18) $5(2 - 2x) \ge -30$

19) $-(7 + 6x) < 29$

20) $12 - 8x \ge -20$

21) $-4(x - 6) > 24$

22) $\dfrac{3x + 9}{6} \le 10$

23) $\dfrac{4x - 10}{3} \le 2$

24) $\dfrac{2x - 8}{3} > 2$

25) $8 + \dfrac{x}{6} < 9$

26) $\dfrac{9x}{7} - 4 < 5$

27) $\dfrac{15x + 45}{15} > 1$

28) $16 + \dfrac{x}{4} < 6$

Solving Systems of Equations by Substitution

✎ Solve each system of equation by substitution.

1) $-x + 3y = -6$

$$x - 2y = 7$$

2) $4x + 2y = -4$

$$-x - 2y = 5$$

3) $2x + 3y = -4$

$$3x - y = 8$$

4) $y = -2x + 3$

$$2x - y = -5$$

5) $4x = 8$

$$5y = 3x + 4$$

6) $4x + 3y = 5$

$$2x + y = -8$$

7) $3x + 2y = 1$

$$x + y = -6$$

8) $3y = x + 2$

$$2x - y = -4$$

Solving Systems of Equations by Elimination

✍ **Solve each system of equation by elimination.**

1) $-4x + y = -4$

$$-y = -3x + 3$$

2) $-x - 3y = -1$

$$x - 2y = 6$$

3) $2x - y = 5$

$$-x + 3y = -5$$

4) $x - y = -12$

$$-2x - 3y = 4$$

5) $7x + 2y = -15$

$$3x - y = -12$$

6) $2x - 3y = -8$

$$x + 2y = 10$$

7) $-3x + 15y = -3$

$$2x - 3y = 9$$

8) $3x - 4y = -6$

$$6x + 5y = -12$$

Systems of Equations Word Problems

✎ **Find the answer for each word problem.**

1) Tickets to a movie cost $4 for adults and $3 for students. A group of friends purchased 8 tickets for $31.00. How many adults ticket did they buy? ____

2) At a store, Eva bought two shirts and five hats for $77.00. Nicole bought three same shirts and four same hats for $84.00. What is the price of each shirt? _____

3) A farmhouse shelters 18 animals, some are pigs, and some are ducks. Altogether there are 66 legs. How many pigs are there? _____

4) A class of 214 students went on a field trip. They took 36 vehicles, some cars and some buses. If each car holds 5 students and each bus hold 22 students, how many buses did they take? _____

5) A theater is selling tickets for a performance. Mr. Smith purchased 5 senior tickets and 3 child tickets for $105 for his friends and family. Mr. Jackson purchased 3 senior tickets and 5 child tickets for $79. What is the price of a senior ticket? $_____

6) The difference of two numbers is 10. Their sum is 20. What is the bigger number? $_____

7) The sum of the digits of a certain two–digit number is 7. Reversing its digits increase the number by 9. What is the number? _____

8) The difference of two numbers is 11. Their sum is 25. What are the numbers? _____

9) The length of a rectangle is 5 meters greater than 2 times the width. The perimeter of rectangle is 28 meters. What is the length of the rectangle? _____

10) Jim has 25 nickels and dimes totaling $1.80. How many nickels does he have? _____

Finding Midpoint

✍ **Find the midpoint of the line segment with the given endpoints.**

1) $(-4, -3), (2, 3)$

2) $(9, 0), (-1, 8)$

3) $(9, -6), (3, 14)$

4) $(-10, -6), (0, 8)$

5) $(2, -5), (14, -15)$

6) $(-10, -3), (4, -13)$

7) $(8, 7), (-8, 13)$

8) $(-3, 6), (-9, 2)$

9) $(-4, 5), (16, -9)$

10) $(7, 14), (9, -2)$

11) $(-8, 6), (6, 6)$

12) $(10, 5), (-2, -3)$

13) $(-5, 12), (-3, 3)$

14) $(12, 7), (8, -2)$

15) $(10, 2), (-6, 14)$

16) $(-1, -2), (-7, 10)$

17) $(7, -7), (13, -13)$

18) $(-3, -8), (11, -4)$

19) $(5, -11), (-8, 9)$

20) $(14, -4), (16, 14)$

21) $(0, -5), (8, -1)$

22) $(3, 0), (-21, 18)$

23) $(17, -3), (-7, -5)$

24) $(26, -12), (6, 24)$

✍ **Find the answer for each problem.**

25) One endpoint of a line segment is $(-3, 7)$ and the midpoint of the line segment is $(-6, 9)$. What is the other endpoint? _____

26) One endpoint of a line segment is $(-3, 7)$ and the midpoint of the line segment is $(1, 5)$. What is the other endpoint? _____

27) One endpoint of a line segment is $(-10, -16)$ and the midpoint of the line segment is $(2, 9)$. What is the other endpoint? _____

Finding Distance of Two Points

✎ **Find the distance between each pair of points.**

1) $(6, 3), (-3, -9)$

2) $(5, 2), (-10, -6)$

3) $(8, 5), (8, 3)$

4) $(-8, -2), (2, 22)$

5) $(6, -7), (-3, -7)$

6) $(12, 0), (-9, -20)$

7) $(3, 20), (3, -5)$

8) $(10, 17), (5, 5)$

9) $(7, -2), (-4, -2)$

10) $(13, 4), (5, -2)$

11) $(11, 13), (5, 5)$

12) $(1, 4), (-23, -3)$

13) $(9, 8), (5, -4)$

14) $(-11, -4), (5, 8)$

15) $(-2, -6), (-2, -12)$

16) $(-1, -4), (23, 3)$

17) $(19, 3), (7, -6)$

18) $(-5, -2), (3, 4)$

19) $(2, 6), (2, -12)$

20) $(-4, -2), (8, -2)$

✎ **Find the answer for each problem.**

21) Triangle ABC is a right triangle on the coordinate system and its vertices are $(-2, 5)$, $(-2, 1)$, and $(1, 1)$. What is the area of triangle ABC? _____

22) Three vertices of a triangle on a coordinate system are $(3, -6)$, $(-5, -12)$, and $(3, -18)$. What is the perimeter of the triangle? _____

23) Four vertices of a rectangle on a coordinate system are $(-2, 2)$, $(-2, 6)$, $(4, 2)$, and $(4, 6)$. What is its perimeter? _____

Answers of Worksheets

One–Step Equations

1) 30	9) 17	17) -14	25) 22
2) 7	10) -4	18) 20	26) 9
3) 4	11) 12	19) 45	27) 60
4) 6	12) 16	20) -13	28) 35
5) 5	13) 34	21) 34	29) 54
6) 11	14) -15	22) -5	30) 24
7) 11	15) -18	23) 47	31) 30
8) 8	16) 14	24) -42	32) -21

Multi–Step Equations

1) 2	9) 2	17) -2	25) 3
2) -7	10) -4	18) -7	26) 14
3) 4	11) -4	19) -5	27) 7
4) 8	12) -5	20) -11	28) 27
5) 6	13) -7	21) -4	29) -5
6) 2	14) 3	22) -2	30) 17
7) 10	15) -2	23) -7	31) -6
8) 2	16) 15	24) 6	32) -4

Graphing Single–Variable Inequalities

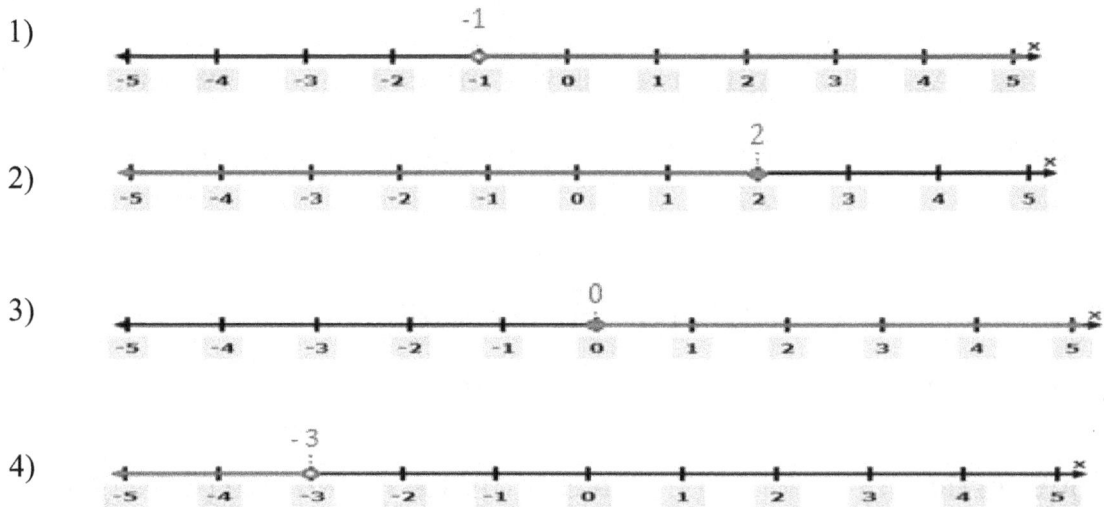

1)

2)

3)

4)

TSI Subject Test Mathematics

5)

6)

7)

8)

One–Step Inequalities

1)

2)

3)

4)

5)

6)

7)

8)

Multi-Step Inequalities

1) $x \le 10$

2) $x \ge 0$

3) $x \le 6$

4) $x \ge 3$

5) $x \ge 8$

6) $x \le 4$

7) $x \le 2$

8) $x \le \frac{23}{6}$

9) $x \le 8$

10) $x \le 7$

11) $x < 10$

12) $x < 3$

13) $x \le 5$

14) $x < 2$

15) $x \ge 11$

16) $x < 2$

17) $x < 3$

18) $x \le 4$

19) $x > -6$

20) $x \le 4$

21) $x < 0$

22) $x \le 17$

23) $x \le 4$

TSI Subject Test Mathematics

24) x > 7

25) x < 6

26) x < 7

27) x > -2

28) x < -40

Solving Systems of Equations by Substitution

1) (9, 1)

2) $(\frac{1}{3}, -\frac{8}{3})$

3) $(\frac{20}{11}, -\frac{28}{11})$

4) $(-\frac{1}{2}, 4)$

5) (2, 2)

6) $(-\frac{29}{2}, 21)$

7) (13, -19)

8) (-2, 0)

Solving Systems of Equations by Elimination

1) (1, 0)

2) (4, -1)

3) (2, -1)

4) (-8, 4)

5) (-3, 3)

6) (2, 4)

7) (6, 1)

8) (-2, 0)

Systems of Equations Word Problems

1) 7

2) $16

3) 15

4) 2

5) $18

6) 15

7) 34

8) 18, 7

9) 11 meters

10) 14

Finding Midpoint

1) (-1, 0)

2) (4, 4)

3) (6, 4)

4) (-5, 1)

5) (8, -10)

6) (-3, -8)

7) (0, 10)

8) (-6, 4)

9) (6, -2)

10) (8, 6)

11) (-1, 6)

12) (4, 1)

13) (-4, 7.5)

14) (10, 2.5)

15) (2, 8)

16) (-4, 4)

17) (10, -10)

18) (4, -6)

19) (-1.5, -1)

20) (15, 5)

21) (4, -3)

22) (-9, 9)

23) (5, -4)

24) (16, 6)

25) (-9, 11)

26) (5, 3)

27) (14, 34)

Finding Distance of Two Points

1) 15

2) 17

3) 2

4) 26

5) 9

6) 29

7) 25

8) 13

9) 11

10) 10

11) 10

12) 25

13) $4\sqrt{10}$

14) 20

15) 6

16) 25

17) 15

18) 10

19) 18

20) 12

21) 6 square units

22) 32 units

23) 20 units

Chapter 9 :

Monomials and Polynomials

Topics that you'll practice in this chapter:

- ✓ GCF of Monomials
- ✓ Factoring Quadratics
- ✓ Factoring by Grouping
- ✓ GCF and Powers of Monomials
- ✓ Writing Polynomials in Standard Form
- ✓ Simplifying Polynomials
- ✓ Adding and Subtracting Polynomials
- ✓ Multiplying Monomials
- ✓ Multiplying and Dividing Monomials
- ✓ Multiplying a Polynomial and a Monomial
- ✓ Multiplying Binomials
- ✓ Factoring Trinomials
- ✓ Operations with Polynomials

Mathematics is, as it were, a sensuous logic, and relates to philosophy as do the arts, music, and plastic art to poetry. — *K. Shegel*

GCF of Monomials

✍ **Find the GCF of each set of monomials.**

1) $26xy, 14x$

2) $40a, 16a^2$

3) $15x^3, 45x^3$

4) $48x^4, 36x^6$

5) $10a^2, 40a^2b$

6) $70a^2, 20a^2b$

7) $32x^2, 24x^2$

8) $55x, 77y^4x$

9) $18, 24x^2, 36$

10) $15v^2, 60v, 45v$

11) p^3q^3, pqn

12) $12m^3n, 18m^3n^3$

13) $15x^3y, 5xy^3z$

14) $44m^2n^3, 66m^4n^5$

15) $28x^4y, 21x^3$

16) $18ab^6, 8a^3b^3c$

17) $25t^2u^3, 35t^4u^2$

18) $16t, 64t^2$

19) $14r^5q, 28pr^3t^2$

20) $20ab^4, 60a^3b$

21) $12d, 42ab^3$

22) $16a^2b^2, 24bc$

23) $10db, 35db$

24) $21m^2n^3, 42m^3n^2$

25) $8xyz, 4x^2$

26) $6x^2yz^{10}, 2x^2yz^4$

27) $120xz, 120y^3, 30y^3$

28) $32b, 40b, 32cb^2$

29) $20x^4, 15x^4, 25x$

30) $110b, 80bc, 70b$

Factoring Quadratics

✍ **Factor each completely.**

1) $x^2 - 13x + 40 =$

2) $m^2 - 11m + 18 =$

3) $p^2 + 2p - 48 =$

4) $3b^2 + 29b + 40 =$

5) $a^2 - 4a + 3 =$

6) $a^2 - 7a + 10 =$

7) $5n^2 - 28n + 32 =$

8) $t^2 - 11t + 28 =$

9) $4x^3 - 27x^2 + 18x =$

10) $x^2 - 10x + 21 =$

11) $7r^2 - 37r + 36 =$

12) $4n^2b - 37nb + 40b =$

13) $3x^2 - 44x + 96 =$

14) $b^3 - 12b^2 + 27b =$

15) $5m^2 - 48m + 27 =$

16) $2x^3 - 23x^2 + 56x =$

17) $x^2 - 13x + 42 =$

18) $p^2 - 13p + 12 =$

19) $x^2 - 14x + 24 =$

20) $8x^2 - 59x + 21 =$

21) $12n^2 - 7n - 10 =$

22) $-3x^2 - 28x - 49 =$

23) $-8x^2 - 2x + 3 =$

24) $6x^2 + 13x - 63 =$

25) $12x^2 - 68x + 40 =$

26) $7x^2 - 48x + 36 =$

27) $8n^2 + 76n + 96 =$

28) $5x^2 - 18x + 9 =$

29) $5x^2 - 35xy =$

30) $-10x^3 - 124x^2y - 48y^2x =$

31) $16a^2 + 8ab - 15b^2 =$

32) $25x^2 - 10xy - 8y^2 =$

33) $64x^2y - 152xy^2 + 90y^3 =$

34) $3x^2 - 12xy - 63y^2 =$

35) $56mp^2 - 49mp =$

36) $27b^2 + 150b + 75 =$

37) $12x^2 + 90xy + 162y^2 =$

38) $16x^2 + 14xy =$

Factoring by Grouping

✎ **Factor each completely.**

1) $18xy - 24x + 3ky - 4k =$

2) $12xy - 10x + 6ny - 5n =$

3) $15n^3 + 10n^2 + 6n + 4 =$

4) $9u^2v + 36u^4 - 6umv - 24u^3m =$

5) $24n^4 + 8n^3 + 36n^2 + 12n =$

6) $16uv - 8u^2 + 24bv - 12bu =$

7) $2x^3 + 6x^2 + 9x + 27 =$

8) $4x^3 + 16x^2 + 8x + 32 =$

9) $3m^3 - 3m^2 + 6m - 6 =$

10) $3x^3 - 9x^2 - 18x + 54 =$

11) $6p^3 + 8p^2 - 15p - 20 =$

12) $18mc + 8md - 9n^2c - 4n^2d =$

13) $16x^4 + 32x^2 - 40x^2 - 80x =$

14) $8xw + 10kx + 12yw + 15ky =$

15) $25xy - 10x + 15ry - 6r =$

16) $3xy - 6x - 7y + 14 =$

17) $5x^3 - 40x^2 + 2x - 16 =$

18) $18x^3 - 126x^2 + 3x - 21 =$

19) $6x^3 + 21x^2 + 10x + 35 =$

20) $12x^3 + 36x^2 - 30x - 90 =$

GCF and Powers of Monomials

✍ **Find the GCF of each pairs of expressions.**

1) $45x^4, 18x^4$

2) $66y, 88y^2x$

3) $18x^2, 24, 36$

4) $20a^3, 55a^2, 15a$

5) xp^3q^3, pq

6) $10m^2n, 45m^2n^2$

7) $15yz, 5xy^2$

8) $24m^8n^3, 16m^2n^2$

9) $18x^4, 27x^2y$

10) $12cb^5, 6a^2b^2c$

11) $15t^2u^6, 20t^4u^7$

12) $17t, 51t^6$

13) $19r^2tq, 38r^4t^9$

14) $12a^2b^9, 36a^8b^5$

15) $14f, 25ab^3$

16) $15a^3b^3c^2, 22abc$

17) $10ab, 5ab$

18) $32m^6n^2, 16m^3n^4$

19) $6xy, 3x^2$

20) $x^2yz^3, 3x^2yz^3$

✍ **Simplify.**

21) $(2x^5)^3$

22) $(2y^33y^2y)^3$

23) $(4x^33x^3)^2$

24) $(5x^2y^4)^4$

25) $(2y^22y^3)^5$

26) $(3x^2y)^7$

27) $(5x^3x^32n)^3$

28) $(4xy^4)^6$

29) $(8x^4y^2)^3$

30) $(14y^5y^2)^2$

31) $(5x^3x^4)^4$

32) $(2x^86x^2k^3)^2$

33) $(y^63y^2)^3$

34) $(2x2x^4)^7$

35) $(8y^6)^2$

36) $(2y^2y^5y^2)^4$

37) $(6y^3y)^4$

38) $(5xy^5)^4$

Writing Polynomials in Standard Form

✎ **Write each polynomial in standard form.**

1) $11x - 7x =$

2) $-5 + 19x - 19x =$

3) $6x^5 - 12x^3 =$

4) $12 + 17x^4 - 12 =$

5) $5x^2 + 4x - 9x^3 =$

6) $-3x^2 + 12x^5 =$

7) $5x + 8x^3 - 2x^8 =$

8) $-7x^3 + 4x - 9x^6 =$

9) $3x^2 + 22 - 6x =$

10) $3 - 4x + 9x^4 =$

11) $13x^2 + 28x - 8x^3 =$

12) $16 + 4x^2 - 2x^3 =$

13) $19x^2 - 9x + 9x^4 =$

14) $3x^4 - 7x^2 - 2x^3 =$

15) $-51 + 3x^2 - 8x^4 =$

16) $7x^2 - 8x^6 + 4x^4 - 15 =$

17) $6x^4 - 4x^5 + 16 - 3x^3 =$

18) $-2x^6 + 4x - 7x^2 - 5x =$

19) $11x^7 + 8x^5 - 5x^7 - 3x^2 =$

20) $2x^2 - 12x^5 + 8x^2 + 3x^6 =$

21) $4x^5 - 11x^7 - 6x^3 + 16x^5 =$

22) $6x^3 + 3x^5 + 34x^4 - 8x^5 =$

23) $3x(4x + 5 - 2x^2) =$

24) $12x(x^6 + 4x^3) =$

25) $5x(3x^2 + 6x + 4) =$

26) $7x(4 - 2x + 6x^5) =$

27) $3x(4x^4 - 4x^3 + 2) =$

28) $4x(2x^5 + 6x^2 - 3) =$

29) $5x(3x^4 + 4x^3 + 2x) =$

30) $2x(3x - 2x^3 + 4x^6) =$

Simplifying Polynomials

✍ **Simplify each expression.**

1) $3(4x - 20) =$

2) $5x(3x - 4) =$

3) $6x(5x - 7) =$

4) $3x(7x + 5) =$

5) $5x(4x - 3) =$

6) $6x(8x + 2) =$

7) $(3x - 2)(x - 4) =$

8) $(x - 5)(2x + 6) =$

9) $(x - 3)(x - 7) =$

10) $(3x + 4)(3x - 4) =$

11) $(5x - 4)(5x - 2) =$

12) $6x^2 + 6x^2 - 8x^4 =$

13) $3x - 2x^2 + 5x^3 + 7 =$

14) $7x + 4x^2 - 10x^3 =$

15) $12x^2 + 5x^5 - 6x^3 =$

16) $-5x^2 + 4x^6 + 6x^8 =$

17) $-12x^3 + 10x^5 - 4x^6 + 4x =$

18) $11 - 7x^2 + 4x^2 - 16x^3 + 11 =$

19) $2x^2 - 9x + 4x^3 + 15x - 10x =$

20) $13 - 7x^5 + 6x^5 - 4x^2 + 5 =$

21) $-5x^8 + x^6 - 14x^3 + 5x^8 =$

22) $(7x^4 - 4) + (7x^4 - 2x^4) =$

23) $3(3x^4 - 4x^3 - 6x^4) =$

24) $-5(x^9 + 8) - 5(10 - x^9) =$

25) $8x^3 - 9x^4 - 2x + 19 - 8x^3 =$

26) $11 - 8x^3 + 6x^3 - 7x^5 + 6 =$

27) $(5x^3 - 4x) - (6x - 2 - 6x^3) =$

28) $4x^2 - 5x^4 - x(3x^3 + 2x) =$

29) $6x + 6x^5 - 10 - 4(x^5 - 3) =$

30) $4 - 3x^4 + (6x^5 - 2x^4 + 5x^5) =$

31) $-(x^5 + 4) - 8(3 + x^5) =$

32) $(4x^3 - 3x) - (3x - 5x^3) =$

Adding and Subtracting Polynomials

✏ **Add or subtract expressions.**

1) $(-2x^2 - 3) + (3x^2 + 4) =$

2) $(4x^3 + 6) - (7 - 2x^3) =$

3) $(4x^5 + 5x^2) - (2x^5 + 15) =$

4) $(6x^3 - 2x^2) + (5x^2 - 4x) =$

5) $(10x^4 + 28x) - (34x^4 + 6) =$

6) $(7x^2 - 3) + (7x^2 + 3) =$

7) $(9x^2 + 4) - (10 - 5x^2) =$

8) $(6x^2 + x^5) - (x^5 + 4) =$

9) $(4x^3 - x) + (3x - 7x^3) =$

10) $(11x + 10) - (8x + 10) =$

11) $(15x^3 - 3x) - (3x - 4x^3) =$

12) $(4x - x^5) - (6x^5 + 8x) =$

13) $(2x^2 - 7x^7) - (4x^7 - 6x) =$

14) $(3x^2 - 5) + (8x^2 + 4x^5) =$

15) $(9x^4 + 5x^5) - (x^5 - 9x^4) =$

16) $(-4x^3 - 2x) + (9x - 5x^3) =$

17) $(4x - 3x^2) - (148x^2 + x) =$

18) $(5x - 8x^4) - (3x^4 - 4x^2) =$

19) $(8x^4 - 4) + (2x^4 - 3x^2) =$

20) $(5x^6 + 7x^3) - (x^3 - 5x^6) =$

21) $(-2x^2 + 20x^5 + 5x^4) + (12x^4 + 8x^5 + 24x^2) =$

22) $(7x^4 - 9x^7 - 6x) - (-3x^4 - 9x^7 + 6x) =$

23) $(14x + 12x^4 - 18x^6) + (20x^4 + 18x^6 - 10x) =$

24) $(5x^8 - 6x^6 - 4x) - (5x^3 + 9x^6 - 7x) =$

25) $(11x^2 - 6x^4 - 3x) - (-4x^2 - 12x^4 + 9x) =$

26) $(-5x^9 + 14x^3 + 3x^7) + (10x^7 + 26x^3 + 3x^9) =$

Multiplying Monomials

✎ **Simplify each expression.**

1) $6u^8 \times (-u^2) =$

2) $(-5p^8) \times (-2p^3) =$

3) $4xy^3z^5 \times 3z^4 =$

4) $3u^5t \times 8ut^4 =$

5) $(-5a^2) \times (-7a^3b^6) =$

6) $-3a^4b^3 \times 6a^2b =$

7) $13xy^5 \times x^4y^4 =$

8) $6p^4q^3 \times (-8pq^6) =$

9) $8s^4t^3 \times 4st^3 =$

10) $(-6x^4y^3) \times 6x^2y =$

11) $3xy^7z \times 12z^3 =$

12) $24xy \times x^2y =$

13) $13pq^4 \times (-3p^2q) =$

14) $13s^3t^4 \times st^4 =$

15) $11p^5 \times (-6p^3) =$

16) $(-8p^3q^5r) \times 3pq^4r^6 =$

17) $(-4a^4) \times (-7a^3b) =$

18) $6u^6v^2 \times (-5u^3v^4) =$

19) $9u^5 \times (-3u) =$

20) $-6xy^5 \times 4x^2y =$

21) $13y^5z^3 \times (-y^3z) =$

22) $8a^4bc^3 \times 2abc^3 =$

23) $(-7p^5q^6) \times (-5p^4q^2) =$

24) $4u^5v^3 \times (-4u^7v^3) =$

25) $17y^4z^5 \times (-y^6z) =$

26) $(-5pq^3r^2) \times 8p^2q^4r =$

27) $3ab^5c^6 \times 5a^4bc^2 =$

28) $6x^3yz^2 \times 3x^2y^7z^3 =$

Multiplying and Dividing Monomials

✎ **Simplify each expression.**

1) $(5x^5)(2x^2) =$

2) $(4x^4)(6x^2) =$

3) $(3x^4)(7x^4) =$

4) $(5x^6)(4x^2) =$

5) $(12x^4)(3x^6) =$

6) $(4yx^8)(8y^4x^3) =$

7) $(14x^4y)(x^3y^5) =$

8) $(-5x^3y^4)(2x^3y^5) =$

9) $(-6x^4y^2)(-3x^3y^5) =$

10) $(5x^3y)(-5x^2y^3) =$

11) $(6x^4y^3)(4x^3y^4) =$

12) $(4x^3y^2)(5x^2y^4) =$

13) $(12x^3y^6)(4x^4y^{10}) =$

14) $(15x^3y^5)(3x^4y^6) =$

15) $(7x^2y^7)(8x^6y^7) =$

16) $(-3x^3y^8)(7x^9y^4) =$

17) $\dfrac{5x^6y^6}{xy^4} =$

18) $\dfrac{19x^7y^5}{19x^6y} =$

19) $\dfrac{56x^4y^4}{8xy} =$

20) $\dfrac{81x^5y^6}{9x^4y^5} =$

21) $\dfrac{36x^7y^6}{9x^2y^3} =$

22) $\dfrac{48x^9y^7}{4x^4y^6} =$

23) $\dfrac{88x^{18}y^{12}}{11x^8y^9} =$

24) $\dfrac{30x^7y^6}{6x^8y^3} =$

25) $\dfrac{150x^7y^6}{30x^4y^6} =$

26) $\dfrac{-42x^{18}y^{14}}{6x^4y^9} =$

27) $\dfrac{-36x^7y^8}{9x^5y^8} =$

Multiplying a Polynomial and a Monomial

✎ **Find each product.**

1) $x(2x + 4) =$

2) $6(4 - 2x) =$

3) $5x(4x + 2) =$

4) $x(-4x + 5) =$

5) $8x(2x - 2) =$

6) $6(2x - 4y) =$

7) $7x(5x - 5) =$

8) $3x(12x + 2y) =$

9) $4x(x + 6y) =$

10) $11x(3x + 4y) =$

11) $7x(3x + 2) =$

12) $10x(4x - 10y) =$

13) $9x(3x - 2y) =$

14) $7x(x - 4y + 6) =$

15) $8x(2x^2 + 5y^2) =$

16) $12x(2x + 3y) =$

17) $4(2x^4 - 4y^4) =$

18) $4x(-3x^2y + 4y) =$

19) $-4(5x^3 - 2xy + 4) =$

20) $4(x^2 - 5xy - 6) =$

21) $8x(2x^3 - 5xy + 2x) =$

22) $-6x(-2x^3 - 6x + 2xy) =$

23) $3(2x^2 + xy - 9y^2) =$

24) $4x(5x^3 - 3x + 7) =$

25) $6(3x^{22} - 2x - 5) =$

26) $x^2(-2x^3 + 4x + 3) =$

27) $x^2(4x^3 + 10 - 2x) =$

28) $4x^4(3x^3 - 2x + 5) =$

29) $2x^2(4x^4 - 5xy + 7y^3) =$

30) $5x^2(5x^4 - 3x + 9) =$

31) $7x^2(6x^2 + 3x - 6) =$

32) $4x(x^3 - 4xy + 2y^2) =$

Multiplying Binomials

✎ **Find each product.**

1) $(x + 3)(x + 6) =$

2) $(x - 4)(x + 3) =$

3) $(x - 3)(x - 8) =$

4) $(x + 8)(x + 9) =$

5) $(x - 2)(x - 12) =$

6) $(x + 5)(x + 5) =$

7) $(x - 6)(x + 7) =$

8) $(x - 8)(x - 3) =$

9) $(x + 7)(x + 12) =$

10) $(x - 4)(x + 8) =$

11) $(x + 8)(x + 8) =$

12) $(x + 2)(x + 7) =$

13) $(x - 6)(x + 6) =$

14) $(x - 5)(x + 5) =$

15) $(x + 11)(x + 11) =$

16) $(x + 6)(x + 9) =$

17) $(x - 2)(x + 2) =$

18) $(x - 4)(x + 7) =$

19) $(3x + 5)(x + 6) =$

20) $(5x - 6)(4x + 8) =$

21) $(x - 7)(3x + 7) =$

22) $(x - 9)(x - 4) =$

23) $(x - 12)(x + 2) =$

24) $(2x - 4)(5x + 4) =$

25) $(3x - 8)(x + 8) =$

26) $(7x - 2)(6x + 3) =$

27) $(4x + 5)(3x + 5) =$

28) $(7x - 4)(9x + 4) =$

29) $(x + 2)(2x - 8) =$

30) $(5x - 4)(5x + 4) =$

31) $(3x + 2)(3x - 7) =$

32) $(x^2 + 8)(x^2 - 8) =$

Factoring Trinomials

✎ **Factor each trinomial.**

1) $x^2 + 8x + 12 =$

2) $x^2 - 6x + 5 =$

3) $x^2 + 15x + 36 =$

4) $x^2 - 12x + 35 =$

5) $x^2 - 11x + 18 =$

6) $x^2 - 9x + 18 =$

7) $x^2 + 18x + 72 =$

8) $x^2 - x - 72 =$

9) $x^2 + 4x - 21 =$

10) $x^2 - 13x + 22 =$

11) $x^2 + 2x - 24 =$

12) $x^2 - 3x - 40 =$

13) $x^2 - 3x - 70 =$

14) $x^2 + 26x + 169 =$

15) $4x^2 - 7x - 15 =$

16) $x^2 - 14x + 33 =$

17) $10x^2 + 5x - 15 =$

18) $6x^2 - 4x - 42 =$

19) $x^2 + 12x + 36 =$

20) $5x^2 + 17x - 12 =$

✎ **Calculate each problem.**

21) The area of a rectangle is $x^2 - x - 56$. If the width of rectangle is $x + 7$, what is its length? _____

22) The area of a parallelogram is $4x^2 + 17x - 15$ and its height is $x + 5$. What is the base of the parallelogram? _____

23) The area of a rectangle is $6x^2 - 22x + 12$. If the width of the rectangle is $3x - 2$, what is its length? _____

Operations with Polynomials

✎ **Find each product.**

1) $4(5x + 3) =$ _____

2) $8(2x + 6) =$ _____

3) $2(5x - 2) =$ _____

4) $-4(7x - 3) =$ _____

5) $3x^2(9x + 1) =$ _____

6) $4x^6(7x - 9) =$ _____

7) $3x^4(-7x + 3) =$ _____

8) $-8x^4(5x - 8) =$ _____

9) $7(x^2 + 5x - 3) =$ _____

10) $9(5x^2 - 7x + 5) =$ _____

11) $3(3x^2 + 3x + 2) =$ _____

12) $5x(3x^2 + 5x + 8) =$ _____

13) $(5x + 7)(3x - 3) =$ _____

14) $(9x + 3)(3x - 5) =$ _____

15) $(6x + 3)(4x - 2) =$ _____

16) $(7x - 2)(3x + 5) =$ _____

✎ **Calculate each problem.**

17) The measures of two sides of a triangle are $(2x + 5y)$ and $(6x - 3y)$. If the perimeter of the triangle is $(13x + 4y)$, what is the measure of the third side? _____

18) The height of a triangle is $(8x + 5)$ and its base is $(4x - 3)$. What is the area of the triangle? _____

19) One side of a square is $(6x + 2)$. What is the area of the square? _____

20) The length of a rectangle is $(5x - 8y)$ and its width is $(15x + 8y)$. What is the perimeter of the rectangle? _____

21) The side of a cube measures $(x + 2)$. What is the volume of the cube? _____

22) If the perimeter of a rectangle is $(28x + 6y)$ and its width is $(5x + 2y)$, what is the length of the rectangle? _____

Answers of Worksheets

GCF of Monomials

1) $2x$

2) $8a$

3) $15x^3$

4) $12x^4$

5) $10a^2$

6) $10a^2$

7) $8x^2$

8) $11x$

9) 6

10) $15v$

11) pq

12) $6m^3n$

13) $5xy$

14) $22m^2n^2$

15) $7x^3$

16) $2ab^3$

17) $5t^2u^2$

18) $16t$

19) $14r^3$

20) $20ab$

21) 6

22) $8b$

23) $5db$

24) $21m^2n^2$

25) $4x$

26) $2x^2yz^4$

27) 30

28) $8b$

29) $5x$

30) $10a$

Factoring Quadratics

1) $(x-5)(x-8)$

2) $(m-2)(m-9)$

3) $(p+8)(p-6)$

4) $(3b+5)(b+8)$

5) $(a-1)(a-3)$

6) $(a-2)(a-5)$

7) $(5n-8)(n-4)$

8) $(t-7)(t-4)$

9) $x(4x-3)(x-6)$

10) $(x-3)(x-7)$

11) $(7r-9)(r-4)$

12) $b(4n-5)(n-8)$

13) $(3x-8)(x-12)$

14) $b(b-3)(b-9)$

15) $(5m-3)(m-9)$

16) $x(2x-7)(x-8)$

17) $(x-6)(x-7)$

18) $(p-1)(p-12)$

19) $(x-2)(x-12)$

20) $(8x-3)(x-7)$

21) $(4n-5)(3n+2)$

22) $-(3x+7)(x+7)$

23) $-(4x+3)(2x-1)$

24) $(2x+9)(3x-7)$

25) $4(x-5)(3x-2)$

26) $(7x-6)(x-6)$

27) $4(2n+3)(n+8)$

28) $(5x-3)(x-3)$

29) $5x(x-7y)$

30) $-2x(5x+2y)(x+12y)$

31) $(4a+5b)(4a-3b)$

32) $(5x+2y)(5x-4y)$

33) $2y(8x-9y)(4x-5y)$

34) $3(x-7y)(x+3y)$

35) $7mp(8p-7)$

36) $3(9b+5)(b+5)$

37) $6(2x + 9y)(x + 3y)$

38) $2x(8x + 7y)$

Factoring by Grouping

1) $(6x + k)(3y - 4)$

2) $(2x + n)(6y - 5)$

3) $(5n^2 + 2)(3n + 2)$

4) $3u(3u - 2m)(v + 4u^2)$

5) $4n(2n^2 + 3)(3n + 1)$

6) $4(2u + 3b)(2v - u)$

7) $(2x^2 + 9)(x + 3)$

8) $4(x^2 + 2)(x + 4)$

9) $3(m^2 + 2)(m - 1)$

10) $3(x^2 - 6)(x - 3)$

11) $(2p^2 - 5)(3p + 4)$

12) $(2m - n^2)(9c + 4d)$

13) $8x(2x^2 - 5)(x + 2)$

14) $(2x + 3y)(4w + 5k)$

15) $(5x + 3r)(5y - 2)$

16) $(3x - 7)(y - 2)$

17) $(5x^2 + 2)(x - 8)$

18) $3(6x^2 + 1)(x - 7)$

19) $(3x^2 + 5)(2x + 7)$

20) $6(2x^2 - 5)(x + 3)$

GCF and Powers of monomials

1) $9x^4$

2) $22y$

3) 6

4) $5a$

5) pq

6) $5m^2n$

7) $5y$

8) $8m^2n^2$

9) $9x^2$

10) $6cb^2$

11) $5t^2u^6$

12) $17t$

13) $19r^2t$

14) $12a^2b^5$

15) 1

16) abc

17) $5ab$

18) $16m^3n^4$

19) $3x$

20) x^2yz^3

21) $8x^{15}$

22) $216y^{18}$

23) $144y^{12}$

24) $625x^8y^{16}$

25) $1,024y^{25}$

26) $2,187x^{14}y^7$

27) $1,000n^3x^{18}$

28) $4,096x^6y^{24}$

29) $512x^{12}y^6$

30) $196y^{14}$

31) $625x^{28}$

32) $144x^{20}k^6$

33) $27y^{24}$

34) $16,384x^{35}$

35) $64y^{12}$

36) $16x^{36}$

37) $1,296y^{36}$

38) $625x^4y^{20}$

Writing Polynomials in Standard Form

1) $4x$

2) -5

3) $6x^5 - 12x^3$

4) $14x^4$

5) $-9x^3 + 5x^2 + 4x$

6) $12x^5 - 3x^2$

7) $-2x^8 + 8x^3 + 5x$

8) $-9x^6 - 7x^3 + 4x$

9) $3x^2 - 6x + 22$

10) $9x^4 - 4x + 3$

11) $-8x^3 + 13x^2 + 28x$

12) $-2x^3 + 4x^2 + 16$

13) $9x^4 + 19x^2 - 9x$

14) $3x^4 - 2x^3 - 7x^2$

15) $-8x^4 + 3x^2 - 51$

16) $-8x^6 + 4x^4 + 7x^2 - 15$

17) $-4x^5 + 6x^4 - 3x^3 + 16$

18) $-2x^6 - 7x^2 - x$

19) $6x^7 + 8x^5 - 3x^2$

20) $3x^6 - 12x^5 + 10x^2$

21) $-11x^7 + 20x^5 - 6x^3$

22) $-5x^5 + 34x^4 + 6x^3$

23) $-6x^3 + 12x^2 + 15x$

24) $12x^7 + 48x^4$

25) $15x^3 + 30x^2 + 20x$

26) $42x^6 - 14x^2 + 28x$

27) $12x^5 - 12x^4 + 6x$

28) $8x^6 + 24x^3 - 12x$

29) $15x^5 + 20x^4 + 10x^2$

30) $8x^7 - 4x^4 + 6x^2$

Simplifying Polynomials

1) $12x - 60$

2) $15x^2 - 20x$

3) $30x^2 - 42x$

4) $21x^2 + 15x$

5) $20x^2 - 15x$

6) $48x^2 + 12x$

7) $3x^2 - 14x + 8$

8) $2x^2 - 4x - 30$

9) $x^2 - 10x + 21$

10) $9x^2 - 16$

11) $25x^2 - 30x + 8$

12) $-8x^4 + 12x^2$

13) $5x^3 - 2x^2 + 3x + 7$

14) $-10x^3 + 4x^2 + 7x$

15) $5x^5 - 6x^3 + 12x^2$

16) $6x^8 + 4x^6 - 5x^2$

17) $-4x^6 + 10x^5 - 12x^3 + 4x$

18) $-16x^3 - 3x^2 + 22$

19) $4x^3 + 2x^2 - 4x$

20) $-x^5 - 4x^2 + 18$

21) $x^6 - 14x^3$

22) $12x^4 - 4$

23) $-9x^4 - 12x^3$

24) -90

25) $-9x^4 - 2x + 19$

26) $-7x^5 - 2x^3 + 17$

27) $11x^3 - 10x + 2$

28) $-8x^4 + 2x^2$

29) $2x^5 + 6x + 2$

30) $11x^5 - 5x^4 + 4$

31) $-9x^5 - 28$

32) $9x^3 - 6x$

TSI Subject Test Mathematics

Adding and Subtracting Polynomials

1) $x^2 + 1$

2) $6x^3 - 1$

3) $2x^5 + 5x^2 - 15$

4) $6x^3 + 3x^2 - 4x$

5) $-24x^4 + 28x - 6$

6) $14x^2$

7) $14x^2 - 6$

8) $6x^2 - 4$

9) $-3x^3 + 2x$

10) $3x$

11) $19x^3 - 6x$

12) $-7x^5 - 4x$

13) $-11x^7 + 2x^2 + 6x$

14) $4x^5 + 11x^2 - 5$

15) $4x^5 + 18x^4$

16) $-9x^3 + 7x$

17) $-151x^2 + 3x$

18) $-11x^4 + 4x^2 + 5x$

19) $10x^4 - 3x^2 - 4$

20) $10x^6 + 6x^3$

21) $28x^5 + 17x^4 + 22x^2$

22) $10x^4 - 12x$

23) $32x^4 + 4x$

24) $5x^8 - 15x^6 - 5x^3 + 3x$

25) $6x^4 + 15x^2 - 12x$

26) $-2x^9 + 13x^7 + 40x^3$

Multiplying Monomials

1) $-6u^{10}$

2) $10p^{11}$

3) $12xy^3z^9$

4) $24u^6t^5$

5) $35a^5b^6$

6) $-18a^6b^4$

7) $13x^5y^9$

8) $-48p^5q^9$

9) $32s^5t^6$

10) $-36x^6y^4$

11) $36xy^7z^4$

12) $24px^3y^2$

13) $-39p^3q^5$

14) $13s^4t^8$

15) $-66p^8$

16) $-24p^4q^9r^7$

17) $28a^7b$

18) $-30u^9v^6$

19) $-27u^6$

20) $-24x^3y^6$

21) $-13y^8z^4$

22) $16a^5b^2c^6$

23) $35p^9q^8$

24) $-16u^{12}v^6$

25) $-17y^{10}z^6$

26) $-40p^3q^7r^3$

27) $15a^5b^6c^8$

28) $18x^5y^8z^5$

Multiplying and Dividing Monomials

1) $10x^7$

2) $24x^6$

3) $21x^8$

4) $20x^8$

5) $36x^{10}$

6) $32x^{11}y^5$

7) $14x^7y^6$

8) $-10x^6y^9$

9) $18x^7y^7$

10) $-25x^5y^4$

11) $24x^7y^7$

12) $20x^5y^6$

13) $48x^7y^{16}$

14) $45x^7y^{11}$

15) $56x^8y^{14}$

16) $-21x^{12}y^{12}$

17) $5x^5y^2$

18) xy^4

19) $7x^3y^3$

20) $9xy$

21) $4x^5y^3$

22) $12x^5y$

23) $8x^{10}y^3$

24) $5x^{-1}y^3$

TSI Subject Test Mathematics

25) $5x^3$ 26) $-7x^{14}y^5$ 27) $-4x^2$

Multiplying a Polynomial and a Monomial

1) $2x^2 + 4x$

2) $-12x + 24$

3) $20x^2 + 10x$

4) $-4x^2 + 5x$

5) $16x^2 - 16x$

6) $12x - 24y$

7) $35x^2 - 35x$

8) $36x^2 + 6xy$

9) $4x^2 + 24xy$

10) $33x^2 + 44xy$

11) $21x^2 + 14x$

12) $40x^2 - 100xy$

13) $27x^2 - 18xy$

14) $7x^2 - 28xy + 42x$

15) $16x^3 + 40xy^2$

16) $24x^2 + 36xy$

17) $8x^4 - 16y^4$

18) $-12x^3y + 16xy$

19) $-20x^3 + 8xy - 16$

20) $4x^2 - 20xy - 24$

21) $16x^4 - 40x^2y + 16x^2$

22) $12x^4 + 36x^2 - 12x^2y$

23) $6x^2 + 3xy - 27y^2$

24) $20x^4 - 12x^2 + 28x$

25) $18x^{22} - 12x - 30$

26) $-2x^5 + 4x^3 + 3x^2$

27) $4x^5 - 2x^3 + 10x^2$

28) $12x^7 - 8x^5 + 20x^4$

29) $8x^6 - 10x^3y + 14x^2y^3$

30) $25x^6 - 15x^3 + 45x^2$

31) $42x^4 + 21x^3 - 42x^2$

32) $4x^4 - 16x^2y + 8xy^2$

Multiplying Binomials

1) $x^2 + 9x + 18$

2) $x^2 - x - 12$

3) $x^2 - 11x + 24$

4) $x^2 + 17x + 72$

5) $x^2 - 14x + 24$

6) $x^2 + 10x + 25$

7) $x^2 + x - 42$

8) $x^2 - 11x + 24$

9) $x^2 + 19x + 84$

10) $x^2 + 4x - 32$

11) $x^2 + 16x + 64$

12) $x^2 + 9x + 14$

13) $x^2 - 36$

14) $x^2 - 25$

15) $x^2 + 22x + 121$

16) $x^2 + 15x + 54$

17) $x^2 - 4$

18) $x^2 + 3x - 28$

19) $3x^2 + 23x + 30$

20) $20x^2 + 16x - 48$

21) $3x^2 - 14x - 49$

22) $x^2 - 13x + 36$

23) $x^2 - 10x - 24$

24) $10x^2 - 12x - 16$

25) $3x^2 + 16x - 64$

26) $42x^2 + 9x - 6$

27) $12x^2 + 35x + 25$

28) $63x^2 - 8x - 16$

29) $2x^2 - 4x - 16$

30) $25x^2 - 16$

31) $9x^2 - 15x - 14$

32) $x^4 - 64$

Factoring Trinomials

1) $(x + 6)(x + 2)$

2) $(x - 5)(x - 1)$

3) $(x + 12)(x + 3)$

4) $(x - 5)(x - 7)$

5) $(x - 2)(x - 9)$

6) $(x - 6)(x - 3)$

7) $(x + 6)(x + 12)$

8) $(x + 8)(x - 9)$

9) $(x - 3)(x + 7)$

10) $(x - 11)(x - 2)$

11) $(x - 4)(x + 6)$

12) $(x - 8)(x + 5)$

13) $(x + 7)(x - 10)$

14) $(x + 13)(x + 13)$

15) $(4x + 5)(x - 3)$

16) $(x - 11)(x - 3)$

17) $(5x - 5)(2x + 3)$

18) $(2x - 6)(3x + 7)$

19) $(x + 6)(x + 6)$

20) $(5x - 3)(x + 4)$

21) $(x - 8)$

22) $(4x - 3)$

23) $(2x - 6)$

Operations with Polynomials

1) $20x + 12$

2) $16x + 48$

3) $10x - 4$

4) $-28x + 12$

5) $27x^3 + 3x^2$

6) $28x^7 - 36x^6$

7) $-21x^5 + 9x^4$

8) $-40x^5 + 64x^4$

9) $7x^2 + 35x - 21$

10) $45x^2 - 63x + 45$

11) $9x^2 + 9x + 6$

12) $15x^3 + 25x^2 + 40x$

13) $15x^2 + 6x - 21$

14) $27x^2 - 36x - 15$

15) $24x^2 - 6$

16) $21x^2 + 29x - 10$

17) $(5x + 2y)$

18) $16x^2 - 2x - \frac{15}{2}$

19) $36x^2 + 24x + 4$

20) $40x$

21) $x^3 + 6x^2 + 12x + 8$

22) $(9x + y)$

Chapter 10 :

Functions Operations and Quadratic

Topics that you'll practice in this chapter:

- ✓ Relations and Functions
- ✓ Evaluating Function
- ✓ Adding and Subtracting Functions
- ✓ Multiplying and Dividing Functions
- ✓ Composition of Functions
- ✓ Quadratic Equation
- ✓ Solving Quadratic Equations
- ✓ Quadratic Formula and the Discriminant
- ✓ Quadratic Inequalities
- ✓ Graphing Quadratic Functions
- ✓ Domain and Range of Radical Functions
- ✓ Solving Radical Equations

It's fine to work on any problem, so long as it generates interesting mathematics along the way – even if you don't solve it at the end of the day." – Andrew Wiles

Relations and Functions

✎ State the domain and range of each relation. Then determine whether each relation is a function.

1)

Function:

....

Domain:

....

Range:

....

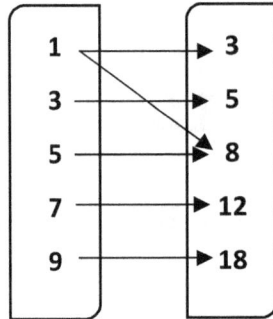

2)

Function:

....

Domain:

....

Range:

....

x	y
3	4
0	1
−2	−3
6	−3
8	2

3)

Function:

....

Domain:

....

Range:

....

4) $\{(1, -2), (4, -1), (0, 5), (4, 0), (3, 8)\}$

Function:

....

Domain:

....

Range:

....

5)

Function:

....

Domain:

....

Range:

....

6)

Function:

....

Domain:

....

Range:

....

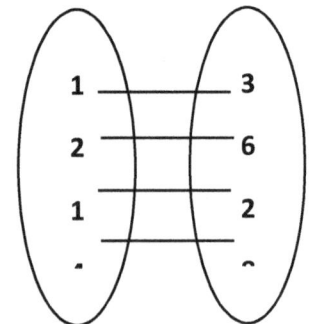

Evaluating Function

✏️ **Write each of following in function notation.**

1) $h = -8x + 3$

2) $k = 2a - 14$

3) $d = 11t$

4) $y = \frac{5}{12}x - \frac{7}{12}$

5) $m = 24n - 210$

6) $c = p^2 - 5p + 10$

✏️ **Evaluate each function.**

7) $f(x) = 2x - 7$, find $f(-3)$

8) $g(x) = \frac{1}{9}x + 12$, find $f(18)$

9) $h(x) = -4x + 9$, find $f(3)$

10) $f(x) = -x + 19$, find $f(-3)$

11) $f(a) = 7a - 12$, find $f(3)$

12) $h(x) = 14 - 3x$, find $f(-4)$

13) $g(n) = 6n - 10$, find $f(2)$

14) $f(x) = -11x - 4$, find $f(-1)$

15) $k(n) = -20 - 3.5n$, find $f(2)$

16) $f(x) = -0.7x + 3.3$, find $f(-7)$

17) $g(n) = \frac{11n+8}{n}$, find $g(2)$

18) $g(n) = \sqrt{3n} + 12$, find $g(3)$

19) $h(x) = x^{-2} - 7$, find $h(\frac{1}{9})$

20) $h(n) = n^{-3} + 11$, find $h(\frac{1}{4})$

21) $h(n) = n^3 - 2$, find $h(\frac{1}{2})$

22) $h(n) = n^2 - 4$, find $h(-\frac{1}{3})$

23) $h(n) = 4n^2 - 13$, find $h(-5)$

24) $h(n) = -2n^3 - 6n$, find $h(2)$

25) $g(n) = \sqrt{16n^2} - \sqrt{n}$, find $g(4)$

26) $h(a) = \frac{-14a+9}{3a}$, find $h(-b)$

27) $k(a) = 12a - 14$, find $k(a - 3)$

28) $h(x) = \frac{1}{9}x + 18$, find $h(-18x)$

29) $h(x) = 8x^2 + 16$, find $h(\frac{x}{2})$

30) $h(x) = x^4 - 20$, find $h(-2x)$

Adding and Subtracting Functions

✎**Perform the indicated operation.**

1) $f(x) = 2x + 3$

   $g(x) = x + 7$

   Find $(f - g)(2)$

2) $g(a) = -5a - 8$

   $f(a) = -3a - 5$

   Find $(g - f)(-2)$

3) $h(t) = 4t + 3$

   $g(t) = 4t + 7$

   Find $(h - g)(t)$

4) $g(a) = -6a - 10$

   $f(a) = 3a^2 + 9$

   Find $(g - f)(x)$

5) $g(x) = \frac{5}{6}x - 23$

   $h(x) = \frac{5}{12}x + 25$

   Find $g(12) - h(12)$

6) $h(x) = \sqrt{3x} - 2$

   $g(x) = \sqrt{3x} + 5$

   Find $(h + g)(12)$

7) $f(x) = x^{-1}$

   $g(x) = x^2 + \frac{5}{x}$

   Find $(f - g)(-3)$

8) $h(n) = n^2 + 2$

   $g(n) = -4n + 6$

   Find $(h - g)(2a)$

9) $g(x) = -2x^2 - 5 - 4x$

   $f(x) = 7 + 2x$

   Find $(g - f)(3x)$

10) $g(t) = 11t - 4$

   $f(t) = -2t^2 + 5$

   Find $(g + f)(-t)$

11) $f(x) = 8x + 9$

   $g(x) = -5x^2 + 3x$

   Find $(f - g)(-x^2)$

12) $f(x) = -3x^4 - 5x$

   $g(x) = 2x^4 + 5x + 22$

   Find $(f + g)(3x^2)$

Multiplying and Dividing Functions

✍ **Perform the indicated operation.**

1) $g(x) = -2x - 1$

$f(x) = 4x + 3$

Find $(g.f)(2)$

2) $f(x) = 5x$

$h(x) = -2x + 3$

Find $(f.h)(-2)$

3) $g(a) = 5a - 2$

$h(a) = 2a - 3$

Find $(g.h)(-3)$

4) $f(x) = 2x - 7$

$h(x) = x - 5$

Find $(\frac{f}{h})(4)$

5) $f(x) = 8a^2$

$g(x) = 3 + 2a$

Find $(\frac{f}{g})(2)$

6) $g(a) = \sqrt{4a} + 2$

$f(a) = (-a)^4 + 1$

Find $(\frac{g}{f})(1)$

7) $g(t) = t^3 + 1$

$h(t) = 5t - 2$

Find $(g.h)(-2)$

8) $g(n) = n^2 + 2n - 4$

$h(n) = -5n + 3$

Find $(g.h)(1)$

9) $g(a) = (a - 3)^2$

$f(a) = a^2 + 4$

Find $(\frac{g}{f})(3)$

10) $g(x) = -3x^2 + \frac{4}{5}x + 9$

$f(x) = x^2 - 24$

Find $(\frac{g}{f})(5)$

11) $f(x) = 2x^3 - 5x^2 + 1$

$g(x) = 3x - 1$

Find $(f.g)(x)$

12) $f(x) = 5x - 2$

$g(x) = x^3 - 2x$

Find $(f.g)(x^2)$

Composition of Functions

✎ Using $f(x) = 2x - 5$ and $g(x) = -2x$, find:

1) $f(g(2)) =$

4) $g(f(5)) =$

2) $f(g(-1)) =$

5) $f(g(3)) =$

3) $g(f(-4)) =$

6) $g(f(0)) =$

✎ Using $f(x) = -\frac{1}{4}x + \frac{3}{4}$ and $g(x) = 2x^2$, find:

7) $g(f(-2)) =$

10) $f(f(1)) =$

8) $g(f(4)) =$

11) $g(f(-4)) =$

9) $g(g(1)) =$

12) $f(g(x)) =$

✎ Using $f(x) = -2x + 2$ and $g(x) = x + 1$, find:

13) $g(f(1)) =$

16) $f(g(-3)) =$

14) $f(f(0)) =$

17) $g(f(2)) =$

15) $f(g(-1)) =$

18) $f(g(x)) =$

✎ Using $f(x) = \sqrt{x + 9}$ and $g(x) = x - 9$, find:

19) $f(g(9)) =$

22) $f(f(7)) =$

20) $g(f(-9)) =$

23) $g(f(-5)) =$

21) $f(g(4)) =$

24) $g(g(0)) =$

Quadratic Equation

✎ **Multiply.**

1) $(x - 4)(x + 6) =$ _____

2) $(x + 5)(x + 7) =$ _____

3) $(x - 6)(x + 8) =$ _____

4) $(x + 2)(x - 9) =$ _____

5) $(x - 7)(x - 8) =$ _____

6) $(3x + 2)(x - 3) =$ _____

7) $(4x - 3)(x + 2) =$ _____

8) $(4x - 5)(x + 1) =$ _____

9) $(7x + 1)(x - 6) =$ _____

10) $(5x + 1)(3x - 3) =$ _____

✎ **Factor each expression.**

11) $x^2 - 2x - 8 =$ _____

12) $x^2 + 8x + 15 =$ _____

13) $x^2 - 2x - 24 =$ _____

14) $x^2 - 10x + 21 =$ _____

15) $x^2 + 10x + 21 =$ _____

16) $4x^2 + 9x + 5 =$ _____

17) $5x^2 + 13x - 6 =$ _____

18) $5x^2 + 17x - 12 =$ _____

19) $2x^2 + 7x + 5 =$ _____

20) $9x^2 - 21x + 6 =$ _____

✎ **Calculate each equation.**

21) $(x + 6)(x - 3) = 0$

22) $(x + 1)(x + 8) = 0$

23) $(3x + 6)(x + 5) = 0$

24) $(2x - 2)(4x + 8) = 0$

25) $x^2 + x + 10 = 22$

26) $x^2 + 11x + 36 = 12$

27) $2x^2 + 9x + 9 = 5$

28) $x^2 + 3x - 24 = 4$

29) $5x^2 + 5x - 40 = 20$

30) $8x^2 + 8x = 48$

Solving Quadratic Equations

✍ **Solve each equation by factoring or using the quadratic formula.**

1) $(x + 9)(x - 1) = 0$

2) $(x + 7)(x + 6) = 0$

3) $(x - 8)(x + 3) = 0$

4) $(x - 6)(x - 4) = 0$

5) $(x + 2)(x + 12) = 0$

6) $(5x + 4)(x + 7) = 0$

7) $(6x + 1)(4x + 5) = 0$

8) $(2x + 7)(x + 8) = 0$

9) $(x + 6)(3x + 15) = 0$

10) $(12x + 2)(x + 8) = 0$

11) $x^2 = 8x$

12) $x^2 - 16 = 0$

13) $3x^2 + 6 = 9x$

14) $-2x^2 - 8 = 10x$

15) $5x^2 + 40x = 45$

16) $x^2 + 10x = 24$

17) $x^2 + 6x = 16$

18) $x^2 + 9x = -18$

19) $x^2 + 13x = -36$

20) $x^2 + 3x - 15 = 5x$

21) $x^2 + 8x + 7 = -8$

22) $3x^2 - 11x = -9 + x$

23) $10x^2 + 3 = 27x - 15$

24) $7x^2 - 6x + 8 = 8$

25) $2x^2 - 12 = -3x + 2$

26) $10x^2 - 26x - 3 = -15$

27) $3x^2 + 21 = -16x + 5$

28) $x^2 + 15x - 10 = -66$

29) $3x^2 - 8x - 8 = 4 + x$

30) $2x^2 + 6x - 24 = 12$

31) $3x^2 - 33x + 54 = -18$

32) $-10x^2 - 15x - 9 = -9 - 27x^2$

Quadratic Formula and the Discriminant

✍ **Find the value of the discriminant of each quadratic equation.**

1) $3x(x - 8) = 0$

2) $2x^2 + 6x - 4 = 0$

3) $x^2 + 6x + 7 = 0$

4) $x^2 - x + 3 = 0$

5) $x^2 + 4x - 3 = 0$

6) $2x^2 + 6x - 10 = 0$

7) $3x^2 + 7x + 5 = 0$

8) $x^2 - 6x - 4 = 0$

9) $2x^2 + 8x + 3 = 0$

10) $x^2 + 7x - 5 = 0$

11) $5x^2 + 2x - 3 = 0$

12) $-3x^2 - 11x + 4 = 0$

13) $-6x^2 - 12x + 8 = 0$

14) $-x^2 - 9x - 12 = 0$

15) $7x^2 - 6x - 10 = 0$

16) $-4x^2 - 2x + 8 = 0$

17) $5x^2 + 8x - 2 = 0$

18) $6x^2 - 4x = 0$

19) $3x^2 - 5x + 2 = 0$

20) $4x^2 + 9x + 3 = 0$

✍ **Find the discriminant of each quadratic equation then state the number of real and imaginary solutions.**

21) $-4x^2 - 16 = 16x$

22) $20x^2 = 20x - 5$

23) $-11x^2 - 19x = 26$

24) $22x^2 - 4x + 1 = 18x^2$

25) $-11x^2 = -15x + 8$

26) $3x^2 + 6x + 9 = 6$

27) $13x^2 - 5x - 12 = -26$

28) $-8x^2 - 32x - 25 = 7$

Graphing Quadratic Functions

✎ketch the graph of each function. Identify the vertex and axis of symmetry.

1) $y = (x + 3)^2 + 2$

2) $y = (x - 3)^2 - 2$

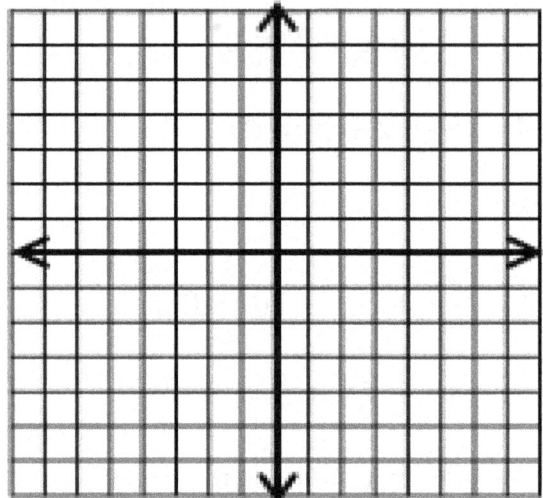

3) $y = 6 - (-x + 4)^2$

4) $y = -3x^2 - 6x + 9$

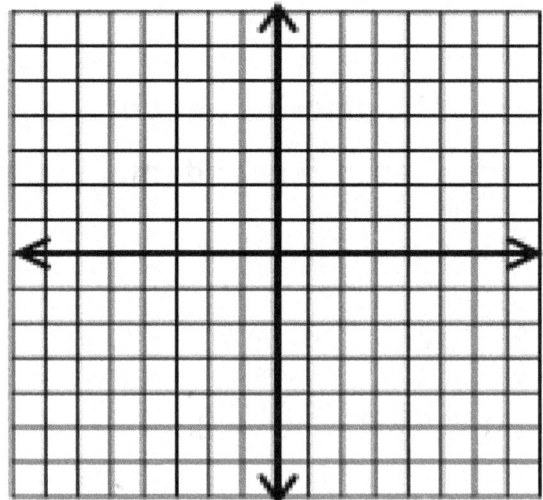

Quadratic Inequalities

✍ **Solve each quadratic inequality.**

1) $x^2 - 25 < 0$

2) $-x^2 - 6x - 8 > 0$

3) $5x^2 + 15x + 30 < 0$

4) $x^2 + 8x + 16 > 0$

5) $2x^2 - 18x - 20 \geq 0$

6) $x^2 > -10x - 25$

7) $3x^2 + 2x + 16 \leq 0$

8) $x^2 - 5x - 14 \leq 0$

9) $x^2 - 6x - 7 \geq 0$

10) $2x^2 + 16x - 18 < 0$

11) $x^2 + 6x - 72 > 0$

12) $3x^2 - 3x - 36 > 0$

13) $x^2 - 15x + 64 \leq 0$

14) $2x^2 - 24x + 72 \leq 0$

15) $x^2 - 16x + 63 \geq 0$

16) $x^2 - 16x + 55 \geq 0$

17) $x^2 - 81 \leq 0$

18) $x^2 - 17x + 42 \geq 0$

19) $9x^2 + 14x + 36 \leq 0$

20) $4x^2 - 2x - 24 > 2x^2$

21) $5x^2 - 20x + 20 < 0$

22) $7x^2 - 6x \geq 6x^2 - 5$

23) $5x^2 - 15 > 4x^2 + 2x$

24) $3x^2 - 4x \geq 3x^2 - 9x + 15$

25) $8x^2 + 9x - 54 > 5x^2$

26) $10x^2 + 50x - 60 < 0$

27) $-x^2 + 15x - 57 \geq 0$

28) $-5x^2 + 25x + 30 \leq 0$

29) $5x^2 + 40x + 75 < 0$

30) $9x^2 + 20x + 180 \leq 0$

31) $3x^2 + 2x - 36 \geq -x$

32) $3x^2 + 9x + 9 \leq 6x^2 + 3x$

Domain and Range of Radical Functions

✍ **Identify the domain and range of each function.**

1) $y = \sqrt{x + 8} - 7$

2) $y = \sqrt[3]{3x - 5} - 4$

3) $y = \sqrt{3x - 9} + 3$

4) $y = \sqrt[3]{(4x + 6)} - 2$

5) $y = 3\sqrt{4x + 20} + 6$

6) $y = \sqrt[3]{(5x - 2)} - 11$

7) $y = 4\sqrt{9x^2 + 8} + 3$

8) $y = \sqrt[3]{(7x^2 - 2)} - 6$

9) $y = 2\sqrt{2x^3 + 16} - 3$

10) $y = \sqrt[3]{(11x + 4)} - 2x$

11) $y = 3\sqrt{-2(4x + 8)} + 5$

12) $y = \sqrt[5]{(3x^2 - 12)} - 6$

13) $y = 3\sqrt{x - 5} - 2$

14) $y = \sqrt[3]{6x + 9} - 4$

✍ **Sketch the graph of each function.**

15) $y = -3\sqrt{x} + 5$

16) $y = 3\sqrt{x} - 6$

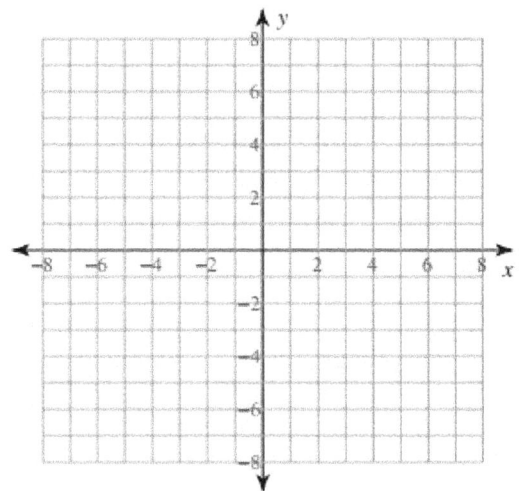

Solving Radical Equations

✎ **Solve each equation. Remember to check for extraneous solutions.**

1) $\sqrt{a} = 9$

2) $\sqrt{v} = 6$

3) $\sqrt{r} = 4$

4) $8 = 16\sqrt{x}$

5) $\sqrt{x+3} = 18$

6) $6 = \sqrt{x-7}$

7) $4 = \sqrt{r-3}$

8) $\sqrt{x-5} = 7$

9) $12 = \sqrt{x-4}$

10) $\sqrt{m+5} = 8$

11) $7\sqrt{5a} = 35$

12) $6\sqrt{2x} = 48$

13) $2 = \sqrt{6x-32}$

14) $\sqrt{304-4x} = 4$

15) $\sqrt{r+2} - 8 = 6$

16) $-21 = -7\sqrt{r+9}$

17) $60 = 6\sqrt{5v}$

18) $x = \sqrt{40-3x}$

19) $\sqrt{90-27a} = 3a$

20) $\sqrt{-8n+88} = 4$

21) $\sqrt{15r-5} = 4r-3$

22) $\sqrt{-64+32x} = 4x$

23) $\sqrt{4x+15} = \sqrt{2x+11}$

24) $\sqrt{12v} = \sqrt{15v-21}$

25) $\sqrt{9-x} = \sqrt{x-3}$

26) $\sqrt{6m+34} = \sqrt{8m+34}$

27) $\sqrt{7r+32} = \sqrt{-8-3r}$

28) $\sqrt{4k+10} = \sqrt{2-4k}$

29) $-20\sqrt{x-13} = -40$

30) $\sqrt{90-2x} = \sqrt{\dfrac{x}{4}}$

Answers of Worksheets

Relation and Functions

1) No, $D_f = \{1, 3, 5, 7, 9\}$, $R_f = \{3, 5, 8, 12, 18\}$

2) Yes, $D_f = \{3, 0, -2, 6, 8\}$, $R_f = \{4, 1, -3, \ 2\}$

3) Yes, $D_f = (-\infty, \infty)$, $R_f = \{2, -\infty)$

4) No, $D_f = \{1, 4, 0, 3\}$, $R_f = \{-2, -1, 5, 0, 8\}$

5) No, $D_f = [-2, \infty)$, $R_f = (-\infty, \infty)$

6) No, $D_f = \{1, 2, 4\}$, $R_f = \{3, 6, 2, 8\}$

Evaluating Function

1) $h(x) = -8x + 3$

2) $k(a) = 2a - 14$

3) $d(t) = 11t$

4) $f(x) = \frac{5}{12}x - \frac{7}{12}$

5) $m(n) = 24n - 210$

6) $c(p) = p^2 - 5p + 10$

7) -13

8) 14

9) -3

10) 22

11) 9

12) 26

13) 2

14) 7

15) -27

16) 8.2

17) 15

18) 15

19) 74

20) 75

21) $-1\frac{7}{8}$

22) $-3\frac{8}{9}$

23) 87

24) -28

25) 14

26) $-\frac{14b+9}{3b}$

27) $12a - 50$

28) $-2x + 18$

29) $2x^2 + 16$

30) $16x^4 - 20$

Adding and Subtracting Functions

1) -2

2) 1

3) -4

4) $-3x^2 - 6x - 19$

5) -43

6) 15

7) $-7\frac{2}{3}$

8) $4a^2 + 8a - 4$

9) $-18x^2 - 18x - 12$

10) $-2t^2 - 11t + 1$

11) $5x^4 - 5x^2 + 9$

12) $-81x^8 + 22$

Multiplying and Dividing Functions

1) -55

2) -70

3) 153

4) -1

5) $4\frac{4}{7}$

6) 2

7) 84

8) 2

9) 0

10) -62

11) $6x^4 - 17x^3 + 5x^2 + 3x - 1$ 12) $5x^8 - 2x^6 - 10x^4 + 4x^2$

Composition of Functions

1) -13
2) -1
3) 26
4) -10
5) -17
6) 10
7) $\frac{25}{8}$

8) $\frac{1}{8}$
9) 8
10) $\frac{5}{8}$
11) $\frac{49}{8}$
12) $-\frac{1}{2}(x^2 - \frac{3}{2})$
13) 1

14) -2
15) 2
16) 6
17) -1
18) $-2x$
19) 3
20) -9

21) 2
22) $\sqrt{13}$
23) -7
24) -18

Quadratic Equations

1) $x^2 + 2x - 24$
2) $x^2 + 12x + 35$
3) $x^2 + 2x - 48$
4) $x^2 - 7x - 18$
5) $x^2 - 15x + 56$
6) $3x^2 - 7x - 6$
7) $4x^2 + 5x - 6$
8) $4x^2 - x - 5$
9) $7x^2 - 41x - 6$
10) $15x^2 - 12x - 3$

11) $(x - 4)(x + 2)$
12) $(x + 5)(x + 3)$
13) $(x - 6)(x + 4)$
14) $(x - 3)(x - 7)$
15) $(x + 3)(x + 7)$
16) $(4x + 5)(x + 1)$
17) $(5x - 2)(x + 3)$
18) $(5x - 3)(x + 4)$
19) $(2x + 5)(x + 1)$
20) $3(x - 2)(3x - 1)$

21) $x = -6, x = 3$
22) $x = -1, x = -8$
23) $x = -2, x = -5$
24) $x = 1, x = -2$
25) $x = 3, x = -4$
26) $x = -3, x = -8$
27) $x = -4, x = -\frac{1}{2}$
28) $x = 4, x = -7$
29) $x = 3, x = -4$
30) $x = -3, x = 2$

Solving quadratic equations

1) $\{-9, 1\}$
2) $\{-6, -7\}$
3) $\{8, -3\}$
4) $\{6, 4\}$
5) $\{-2, -12\}$
6) $\{-\frac{4}{5}, -7\}$
7) $\{-\frac{5}{4}, -\frac{1}{6}\}$
8) $\{-\frac{7}{2}, -8\}$

9) $\{-6, -5\}$
10) $\{-\frac{1}{6}, -8\}$
11) $\{8, 0\}$
12) $\{4, -4\}$
13) $\{2, 1\}$
14) $\{-4, -1\}$
15) $\{1, -9\}$
16) $\{2, -12\}$

17) $\{2, -8\}$
18) $\{-3, -6\}$
19) $\{-4, -9\}$
20) $\{5, -3\}$
21) $\{-5, -3\}$
22) $\{1, 3\}$
23) $\{\frac{6}{5}, \frac{3}{2}\}$
24) $\{\frac{6}{7}, 0\}$

25) $\{-\frac{7}{2}, 2\}$
26) $\{\frac{3}{5}, 2\}$
27) $\{-\frac{4}{3}, -4\}$
28) $\{-8, -7\}$
29) $\{4, -1\}$
30) $\{3, -6\}$
31) $\{3, 8\}$
32) $\{\frac{15}{17}, 0\}$

Quadratic formula and the discriminant

1) 576	6) 116	11) 64	16) 132
2) 68	7) −11	12) 169	17) 104
3) 8	8) 52	13) 336	18) 16
4) −11	9) 40	14) 33	19) 1
5) 28	10) 69	15) 316	20) 33

21) 0, *one real solution*   24) 0, *one real solution*   27) -703, *no solution*

22) 0, *one real solution*   25) -127, *no solution*   28) 0, *one real solution*

23) -783, *no solution*   26) 0, *one real solution*

Graphing quadratic functions

1) $(-3, 2), x = -3$

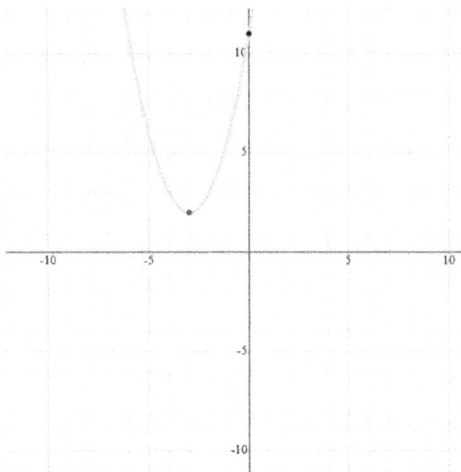

2) $(3, -2), x = 3$

3) $(4, 6), x = 4$

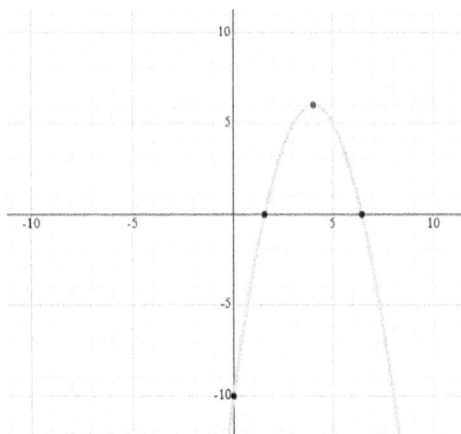

4) $(-1, 12), x = -1$

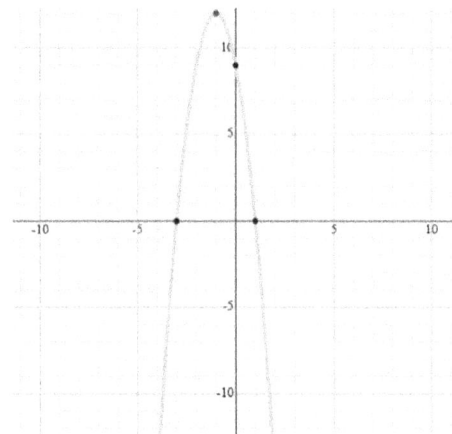

Quadratic inequalities

1) $-5 < x < 5$

2) $-4 < x < -2$

3) no solution

4) $x < -4 \ or \ x > -4$

5) $x \leq -1 \ or \ x \geq 10$

6) $x < -5 \ or \ x > -5$

7) no solution

8) $-2 \leq x \leq 7$

9) $x \leq -1 \ or \ x \geq 7$

10) $-9 < x < 1$

11) $x < -12 \ or \ x > 6$

12) $-3 < x < 4$

13) no solution

14) $x = 6$

15) $x \leq 7 \ or \ x \geq 9$

16) $x \leq 5 \ or \ x \geq 11$

17) $-9 \leq x \leq 9$

18) $x \leq 3 \ or \ x \geq 14$

19) no solution

20) $x < -3 \ or \ x > 4$

21) no solution

22) $x \leq 1 \ or \ x \geq 5$

23) $x < -3 \ or \ x > 5$

24) $x \geq 3$

25) $x < -6 \ or \ x > 3$

26) $-6 < x < 1$

27) no solution

28) $x \leq -1 \ or \ x \geq 6$

29) $-5 < x < -3$

30) no solution

31) $x \leq -4 \ or \ x \geq 3$

32) $x \leq -1 \ or \ x \geq 3$

Domain and range of radical functions

1) domain: $x \geq -8$
   range: $y \geq -7$

2) domain: {all real numbers}
   range: {all real numbers}

3) domain: $x \geq 3$
   range: $y \geq 3$

4) domain: {all real numbers}
   range: {all real numbers}

5) domain: $x \geq -5$
   range: $y \geq 6$

6) domain: {all real numbers}
   range: {all real numbers}

7) domain: {all real numbers}
   range: $y \geq 8\sqrt{2} + 3$

8) domain: {all real numbers}
   range: {all real numbers}

9) domain: $x \geq -2$
   range: $y \geq -3$

10) domain: {all real numbers}
    range: {all real numbers}

11) domain: $x \leq -2$
    range: $y \geq 5$

12) domain: {all real numbers}
    range: {all real numbers}

13) domain: $x \geq 5$
    range: $y \geq -2$

14) domain: {all real numbers}
    range: {all real numbers}

15)

16)

Solving radical equations

1) $\{81\}$	11) $\{5\}$	21) $\{2\}$
2) $\{36\}$	12) $\{32\}$	22) no solution
3) $\{16\}$	13) $\{6\}$	23) $\{-2\}$
4) $\{\frac{1}{4}\}$	14) $\{72\}$	24) $\{7\}$
5) $\{321\}$	15) $\{194\}$	25) $\{6\}$
6) $\{43\}$	16) $\{0\}$	26) $\{0\}$
7) $\{19\}$	17) $\{20\}$	27) $\{-4\}$
8) $\{54\}$	18) $\{5\}$	28) $\{-1\}$
9) $\{148\}$	19) $\{2\}$	29) $\{17\}$
10) $\{59\}$	20) $\{9\}$	30) $\{40\}$

Chapter 11 :

Rational Expressions

Topics that you will learn in this chapter:

✓ Simplifying and Graphing Rational Expressions

✓ Adding and Subtracting Rational Expressions

✓ Multiplying and Dividing Rational Expressions

✓ Solving Rational Equations and Complex Fractions

"What music is to the heart; mathematics is to the mind."

— Amit Kalantri

Simplifying and Graphing Rational Expressions

✍ Simplify.

1) $\dfrac{x+3}{3x+9} =$

5) $\dfrac{16x^4}{24x} =$

2) $\dfrac{3x^2+12x+12}{x+2} =$

6) $\dfrac{x-1}{x^2+4x-5} =$

3) $\dfrac{8}{4x-4} =$

7) $\dfrac{x^2-5x-14}{x-7} =$

4) $\dfrac{x^2+6x+5}{x^2+9x+20} =$

8) $\dfrac{36}{6x-6} =$

✍ Identify the points of discontinuity, holes, vertical asymptotes, x-intercepts, and horizontal asymptote of each.

9) $f(x) = \dfrac{x^2-x+2}{-4x^2-4x+8} =$

11) $f(x) = \dfrac{x-2}{x-7} =$

10) $f(x) = \dfrac{-x-2}{-3x^2-15x-18} =$

12) $f(x) = \dfrac{3x^2}{3x^2-3x-6} =$

✍ Graph rational expressions.

13) $f(x) = \dfrac{-x^2+3x-4}{x-3}$

14) $f(x) = \dfrac{-x^3-10x+32}{x^2-x-3}$

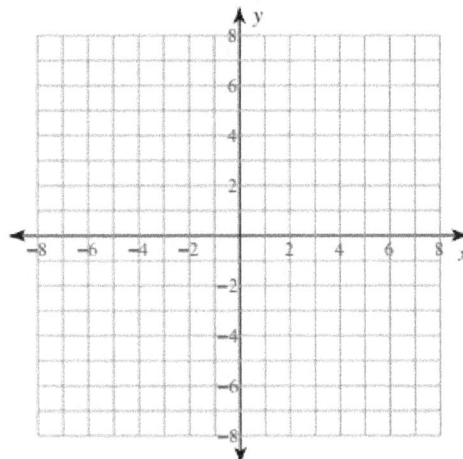

Adding and Subtracting Rational Expressions

✎ **Simplify each expression.**

1) $\dfrac{3}{3x+7} + \dfrac{x-5}{3x+7} =$

10) $\dfrac{3}{x^2 - 2x - 8} + \dfrac{-3}{x^2 - 4} =$

2) $\dfrac{x+2}{x-2} + \dfrac{x-2}{x+5} =$

11) $\dfrac{3}{x+4} - \dfrac{1}{x+2} =$

3) $\dfrac{2}{x+3} - \dfrac{5}{x-8} =$

12) $\dfrac{4x+4}{4x^2 + 12x - 16} + \dfrac{5x}{3x} =$

4) $\dfrac{x-3}{x^2 - 11} - \dfrac{x-4}{11 - x^2} =$

13) $2 + \dfrac{x}{x+2} - \dfrac{2}{x^2 - 4} =$

5) $\dfrac{2}{x+4} + \dfrac{6x}{3x + 12} =$

14) $\dfrac{3}{x+2} - \dfrac{3}{x+5} =$

6) $\dfrac{7+x}{2x} + \dfrac{x-3}{2x} =$

15) $\dfrac{1}{5x^2 + 15x} + \dfrac{3}{2x} =$

7) $3 + \dfrac{x-4}{x+3} =$

16) $\dfrac{x^2 + 4x + 4}{4x + 8} + \dfrac{3x+3}{x+1} =$

8) $\dfrac{3x}{3x+5} + \dfrac{5x}{4x+1} =$

17) $\dfrac{x}{3x+5} + \dfrac{3x}{3x+4} =$

9) $\dfrac{x+y}{y-x} - \dfrac{2xy}{y^2 - x^2} =$

18) $\dfrac{3}{12+4x} - \dfrac{3x-5}{4x^2 + 12x} =$

Multiplying and Dividing Rational Expressions

✍ **Simplify each expression.**

1) $\dfrac{10x}{15} \times \dfrac{15}{12x} =$

2) $\dfrac{63x}{12} \times \dfrac{60}{28x^2} =$

3) $\dfrac{75}{4} \times \dfrac{24x}{87} =$

4) $\dfrac{58}{27} \times \dfrac{27x^2}{43} =$

5) $\dfrac{90}{14x} \times \dfrac{32x}{45x} =$

6) $\dfrac{6x+30}{x+2} \times \dfrac{x+2}{6} =$

7) $\dfrac{x-9}{x+5} \times \dfrac{8x+40}{x-9} =$

8) $\dfrac{2}{x+11} \times \dfrac{5x+55}{5x+10} =$

9) $\dfrac{4(x+5)}{5x} \times \dfrac{9}{4(x+5)} =$

10) $\dfrac{7(x+3)}{x+3} \times \dfrac{7x}{7(x-5)} =$

11) $\dfrac{5x^2+10x}{x+6} \times \dfrac{1}{x+2} =$

12) $\dfrac{16x^2-16x}{14x^2-14x} \times \dfrac{6x}{6x^2} =$

13) $\dfrac{1}{x-7} \times \dfrac{x^2+4x-21}{x+7} =$

14) $\dfrac{x^2-12x+36}{12x-72} \times \dfrac{x-6}{36-6x} =$

✍ **Divide.**

15) $\dfrac{-1+2x-x^2}{x^2+4x-5} \div \dfrac{3x}{x+5} =$

16) $\dfrac{9a}{a+7} \div \dfrac{9a}{2a+14} =$

17) $\dfrac{13x}{x-9} \div \dfrac{13x}{8x-72} =$

18) $\dfrac{3x+12}{8x^2-80x} \div \dfrac{3}{8x} =$

19) $\dfrac{x-3}{x+3x-12} \div \dfrac{11x}{x+8} =$

20) $\dfrac{6x}{x-8} \div \dfrac{6x}{12x-96} =$

21) $\dfrac{x+6}{x^2+14x+48} \div \dfrac{5x}{x+8} =$

22) $\dfrac{x+2}{x^2+15x+54} \div \dfrac{5x}{x+9} =$

23) $\dfrac{16x+14}{4} \div \dfrac{64x+56}{4x} =$

24) $\dfrac{7x^3+49x^2}{x^2+16x+63} \div \dfrac{4(x+6)}{4x^3+24x^2} =$

25) $\dfrac{x^2+9x+14}{x^2+6x+8} \div \dfrac{1}{x+4} =$

26) $\dfrac{x^2+3x-10}{6x+30} \div \dfrac{3}{x+7} =$

27) $\dfrac{x+6}{x^2+10x+21} \div \dfrac{1}{x+7} =$

28) $\dfrac{1}{4x} \div \dfrac{8x}{2x^2+18x} =$

Solving Rational Equations and Complex Fractions

✍ **Solve each equation. Remember to check for extraneous solutions.**

1) $\dfrac{x-1}{x+3} = \dfrac{2x-2}{x-3}$

2) $\dfrac{1}{x} = \dfrac{5}{6x} + 2$

3) $\dfrac{3x-4}{6x+1} = \dfrac{x+4}{x-1}$

4) $\dfrac{1}{6b^2} + \dfrac{1}{6b} = \dfrac{1}{b^2}$

5) $\dfrac{2x-1}{5x+1} = \dfrac{3x-4}{x-7}$

6) $\dfrac{1}{3n^2} - \dfrac{1}{n} = \dfrac{1}{4n^2}$

7) $\dfrac{1}{8b^2} = \dfrac{1}{4b^2} - \dfrac{1}{2b}$

8) $\dfrac{1}{n-4} - 2 = \dfrac{3}{n-4}$

9) $\dfrac{4}{r-2} = -\dfrac{8}{r+2}$

10) $2 = \dfrac{2}{x^2+4x} + \dfrac{2x+2}{x}$

11) $\dfrac{3}{x} = 7 + \dfrac{2}{3x}$

12) $\dfrac{x+3}{x^2-x} - 2 = \dfrac{1}{x^2-x}$

13) $\dfrac{x-2}{x+5} - 1 = \dfrac{1}{x+2}$

14) $\dfrac{1}{12x^2} = \dfrac{1}{6x^2} - \dfrac{1}{2x}$

15) $\dfrac{x+3}{x^2+2x} = \dfrac{x}{x^2+2x} - \dfrac{6}{x+2}$

16) $1 = \dfrac{5}{2x^2+4x} + \dfrac{x+2}{2x}$

✍ **Simplify each expression.**

17) $\dfrac{\frac{3}{5}}{\frac{4}{45} - \frac{5}{18}} =$

18) $\dfrac{\frac{17}{2}}{-7\frac{4}{15}} =$

19) $\dfrac{7}{\frac{7}{x} + \frac{3}{4x}} =$

20) $\dfrac{2x^2}{\frac{3}{8} - \frac{3}{x}} =$

21) $\dfrac{\frac{2}{x-1} - \frac{1}{x+4}}{\frac{3}{x^2+9x+20}} =$

22) $\dfrac{\frac{14}{x-1}}{\frac{14}{7} - \frac{14}{42}} =$

23) $\dfrac{1 + \frac{8}{x-4}}{1 - \frac{4}{x-4}} =$

24) $\dfrac{\frac{1}{3} - \frac{x+2}{6}}{\frac{x^2}{4} - \frac{2}{3}} =$

Answers of Worksheets

Simplifying and Graphing rational expressions

1) $\frac{1}{3}$

2) $3(x+2)$

3) $\frac{2}{x-1}$

4) $\frac{x+1}{x+4}$

5) $\frac{2x^3}{3}$

6) $\frac{1}{x+5}$

7) $x+2$

8) $\frac{6}{x-1}$

9) Discontinuities: –2, 1; Vertical Asymptote: $x=-2, x=1$; Holes: None
Horizontal. Asymptote: $y=-\frac{1}{4}$; x–intercepts: None

10) Discontinuities –2, –3; Vertical Asymptote $x=-3$; Holes $x=-2$
Horizontal Asymptote $y=0$; x–intercepts. None

11) Discontinuities: 7; Vertical Asymptote: $x=7$; Holes: None
Horizontal Asymptote: $y=1$; x–intercepts: 2

12) Discontinuities: –1, 2; Vertical Asymptote: $x=-1, x=2$; Holes: None
Horizontal Asymptote: $y=1$; x–intercepts: 0

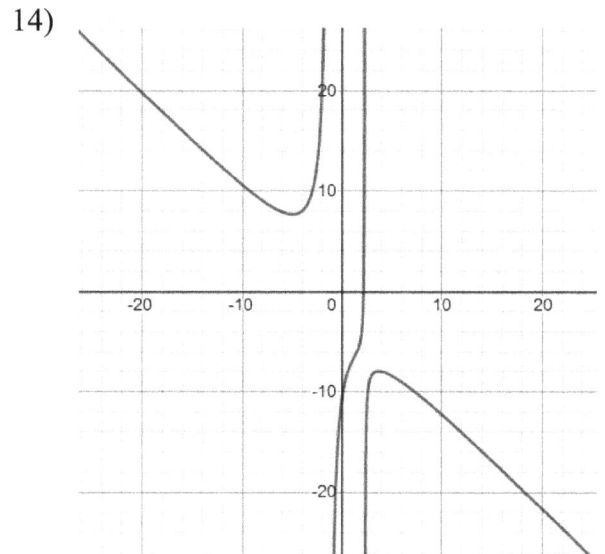

13)

14)

Adding and subtracting rational expressions

1) $\frac{x-2}{3x+7}$

2) $\frac{2x^2+3x+14}{(x-2)(x+5)}$

3) $\frac{-3x-31}{(x+3)(x-8)}$

4) $\frac{2x-7}{x^2-11}$

5) $\frac{2+2x}{x+4}$

6) $\frac{x+2}{x}$

7) $\frac{x-1}{x+3}$

8) $\frac{27x^2+28x}{(3x+5)(4x+1)}$

9) $\frac{x^2+y^2}{(x-y)(x+y)}$

10) $\frac{6}{(x+2)(x-4)(x-2)}$

13) $\frac{3x^2 - 2x - 10}{(x+2)(x-2)}$

16) $\frac{x+2}{4} + 3$

11) $\frac{2x+2}{(x+4)(x+2)}$

14) $\frac{9}{(x+2)(x+5)}$

17) $\frac{12x^2 + 19x}{(3x+5)(3x+4)}$

12) $\frac{5x^2 + 18x - 17}{3(x-1)(x+4)}$

15) $\frac{15x+47}{10x(x+3)}$

18) $\frac{5}{4x(x+3)}$

Multiplying and Dividing rational expressions

1) $\frac{5}{6}$

11) $\frac{5x}{x+6}$

21) $\frac{1}{5x}$

2) $\frac{45}{4x}$

12) $\frac{8}{7x}$

22) $\frac{x+2}{5x(x+6)}$

3) $\frac{150x}{29}$

13) $\frac{x-3}{x-7}$

23) $\frac{x}{4}$

4) $\frac{58x^2}{43}$

14) $-\frac{(x-6)}{72}$

24) $\frac{7x^4}{x+9}$

5) $\frac{32}{7x}$

15) $-\frac{x-1}{3x}$

25) $x+7$

6) $x+5$

16) 2

26) $\frac{(x-2)(x+7)}{18}$

7) 8

17) 8

27) $\frac{x+6}{x+3}$

8) $\frac{2}{x+2}$

18) $\frac{x+4}{x-10}$

28) $\frac{x+9}{16x}$

9) $\frac{9}{5x}$

19) $\frac{x+8}{44x}$

10) $\frac{7x}{x-5}$

20) 12

Solving rational equations and complex fractions

1) $\{1, -9\}$

9) $\{\frac{2}{3}\}$

18) $-1\frac{37}{218}$

2) $\{\frac{1}{12}\}$

10) $\{-5\}$

19) $\frac{28x}{31}$

3) $\{\frac{-32}{3}, 0\}$

11) $\{\frac{1}{3}\}$

20) $\frac{16x^3}{3x-24}$

4) $\{5\}$

12) $\{2, -\frac{1}{2}\}$

21) $\frac{(x+9)(x+5)}{3(x-1)}$

5) $\{-\frac{11}{13}, 1\}$

13) $\{-\frac{19}{8}\}$

22) $\frac{42}{5(x-1)}$

6) $\{\frac{1}{12}\}$

14) $\{\frac{1}{6}\}$

23) $\frac{x+4}{x-8}$

7) $\{\frac{1}{4}\}$

15) $\{-\frac{1}{2}\}$

24) $-\frac{2x}{3x^2-8}$

8) $\{3\}$

16) $\{3, -3\}$

17) $-\frac{54}{17}$

Chapter 12 :

Statistics and Probability

Topics that you'll practice in this chapter:

✓ Mean and Median

✓ Mode and Range

✓ Probability Problems

✓ Factorials

✓ Combinations and Permutations

"The book of nature is written in the language of Mathematic" - *Galileo.*

Mean and Median

✎ **Find Mean and Median of the Given Data.**

1) 8, 7, 14, 4, 8

2) 14, 8, 25, 19, 16, 33, 11

3) 23, 18, 15, 12, 17

4) 34, 14, 10, 15, 6, 11

5) 10, 19, 6, 8, 32, 20, 17

6) 17, 26, 39, 69, 20, 6

7) 40, 38, 18, 11, 9, 2, 7, 32, 41

8) 24, 21, 31, 12, 33, 32, 22

9) 16, 14, 20, 41, 15, 20, 38, 4

10) 20, 20, 30, 18, 6, 28, 12, 46

11) 12, 7, 10, 11, 16, 22

12) 10, 29, 27, 12, 2, 15, 10, 3

✎ **Calculate.**

13) In a javelin throw competition, five athletics score 56, 34, 62, 23 and 19 meters. What are their Mean and Median? _____

14) Eva went to shop and bought 8 apples, 14 peaches, 6 bananas, 4 pineapples and 12 melons. What are the Mean and Median of her purchase? _____

15) Bob has 17 black pen, 19 red pen, 14 green pens, 20 blue pens and 5 boxes of yellow pens. If the Mean and Median are 19 respectively, what is the number of yellow pens in each box? _____

Mode and Range

✍ **Find Mode and Rage of the Given Data.**

1) 4, 3, 7, 3, 3, 4

    Mode: _____   Range: _____

2) 18, 18, 24, 26, 18, 8, 14, 22

    Mode: _____   Range: _____

3) 8, 8, 8, 16, 19, 22, 20, 9, 13

    Mode: _____   Range: _____

4) 24, 24, 14, 28, 20, 18, 20, 24

    Mode: _____   Range: _____

5) 6, 21, 27, 24, 27, 27

    Mode: _____   Range: _____

6) 21, 8, 8, 7, 8, 12, 10, 22, 18, 13

    Mode: _____   Range: _____

7) 7, 4, 4, 6, 13, 13, 13, 0, 2, 2

    Mode: _____   Range: _____

8) 5, 8, 5, 14, 12, 14, 3, 5, 18

    Mode: _____   Range: _____

9) 7, 7, 7, 12, 7, 3, 8, 16, 3, 17

    Mode: _____   Range: _____

10) 15, 15, 19, 16, 4, 16, 10, 15

    Mode: _____   Range: _____

11) 6, 6, 5, 6, 42, 13, 19, 2

    Mode: _____   Range: _____

12) 8, 8, 9, 8, 9, 4, 34, 22

    Mode: _____   Range: _____

✍ **Calculate.**

13) A stationery sold 12 pencils, 56 red pens, 24 blue pens, 20 notebooks, 12 erasers, 21 rulers and 11 color pencils. What are the Mode and Range for the stationery sells?

          Mode: _____   Range: _____

14) In an English test, eight students score 10, 15, 15, 18 18, 16, 15 and 15. What are their Mode and Range? _____

15) What is the range of the first 6 even numbers greater than 8?

    _____

Stem–and–Leaf Plot

✎ **Make stem ad leaf plots for the given data.**

1) 24, 26, 29, 20, 53, 27, 51, 55, 36, 21, 37, 30

Stem | Leaf plot

2) 11, 59, 66, 14, 18, 19, 59, 65, 69, 61, 68, 65

Stem | Leaf plot

3) 121, 55, 66, 54, 112, 128, 63, 125, 59, 123, 68, 119

Stem | Leaf plot

4) 51, 32, 100, 56, 84, 36, 107, 56, 85, 39, 56, 106, 89

Stem | Leaf plot

5) 33, 89, 19, 87, 81, 16, 11, 30, 86, 35, 17, 35, 13

Stem | Leaf plot

6) 60, 92, 22, 25, 67, 93, 95, 62, 21, 64, 98, 29

Stem | Leaf plot

Pie Graph

The circle graph below shows all Robert's expenses for last month. Robert spent $140 on his hobbies last month.

Answer following questions based on the Pie graph.

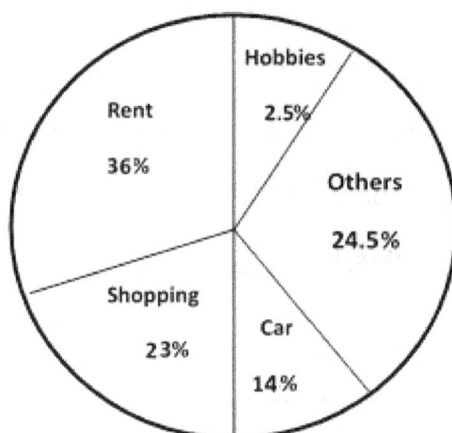

1) How much was Robert's total expenses last month? _____

2) How much did Robert spend on his car last month? _____

3) How much did Robert spend for shopping last month? _____

4) How much did Robert spend on his rent last month? _____

5) What fraction is Robert's expenses for his rent and car out of his total

   expenses last month? _____

Probability Problems

✎ **Calculate.**

1) A number is chosen at random from 1 to 10. Find the probability of selecting number 6 or smaller numbers. _____

2) Bag A contains 18 red marbles and 6 green marbles. Bag B contains 16 black marbles and 8 orange marbles. What is the probability of selecting a green marble at random from bag A? What is the probability of selecting a black marble at random from Bag B? _____

3) A number is chosen at random from 1 to 20. What is the probability of selecting multiples of 4? _____

4) A card is chosen from a well-shuffled deck of 52 cards. What is the probability that the card will be a queen? _____

5) A number is chosen at random from 1 to 15. What is the probability of selecting a multiple of 3 or 5? _____

A spinner numbered 1–8, is spun once. What is the probability of spinning …?

6) an Odd number? _____   7) a multiple of 2? _____

8) a multiple of 5? _____   9) number 10? _____

Factorials

✎ **Determine the value for each expression.**

1) $4! + 0! =$

2) $2! + 5! =$

3) $(2!)^2 =$

4) $5! - 3! =$

5) $6! - 3! + 10 =$

6) $3! \times 4 - 15 =$

7) $(2! + 3!)^2 =$

8) $(4! - 3!)^2 =$

9) $(3! \, 0!)^2 - 10 =$

10) $\dfrac{10!}{8!} =$

11) $\dfrac{6!}{4!} =$

12) $\dfrac{6!}{5!} =$

13) $\dfrac{15!}{13!} =$

14) $\dfrac{n!}{(n-3)!} =$

15) $\dfrac{(n+2)!}{n!} =$

16) $\dfrac{(2+2!)^3}{2!} =$

17) $\dfrac{5(n+2)!}{(n+1)!} =$

18) $\dfrac{22!}{20!4!} =$

19) $\dfrac{13!}{11!3!} =$

20) $\dfrac{9 \times 210!}{3(7 \times 30)!} =$

21) $\dfrac{32!}{31!2!} =$

22) $\dfrac{11!12!}{10!13!} =$

23) $\dfrac{16!15!}{14!14!} =$

24) $\dfrac{(5 \times 3)!}{0!14!} =$

25) $\dfrac{4!(5n-2)!}{(5n)!} =$

26) $\dfrac{4n(4n+7)!}{(4n+8)!} =$

27) $\dfrac{(n-2)!(n+1)}{(n+2)!} =$

Combinations and Permutations

✎ **Calculate the value of each.**

1) 6! = ____

2) 2! × 5! = ____

3) 3 × 4! = ____

4) 5! + 3! = ____

5) 7! = ____

6) 4! = ____

7) 3! + 3! = ____

8) 7! − 5! = ____

✎ **Find the answer for each word problems.**

9) Susan is baking cookies. She uses sugar, butter, Vanilla, eggs and flour. How many different orders of ingredients can she try? _____

10) Albert is planning for his vacation. He wants to go to museum, watch a movie, go to the beach, play the game and play football. How many ways of ordering are there for him? _____

11) How many 4-digit numbers can be named using the digits 3, 4, 5, and 6 without repetition? _____

12) In how many ways can 5 boys be arranged in a straight line? _____

13) In how many ways can 6 athletes be arranged in a straight line? _____

14) A professor is going to arrange her 7 students in a straight line. In how many ways can she do this? _____

15) How many code symbols can be formed with the letters for the word GAMES? _____

16) In how many ways a team of 7 basketball players can choose a captain and co-captain? _____

Answers of Worksheets

Mean and Median

1) Mean: 8.2, Median: 8
2) Mean: 18, Median: 16
3) Mean: 17, Median: 17
4) Mean: 15, Median: 12.5
5) Mean: 16, Median: 17

6) Mean: 29.5, Median: 23
7) Mean: 22, Median: 18
8) Mean: 25, Median: 24
9) Mean: 21, Median: 18
10) Mean: 22.5, Median: 20

11) Mean: 13, Median: 11.5
12) Mean: 13.5, Median: 11
13) Mean: 38.8, Median: 34
14) Mean: 8.8, Median: 8
15) 5

Mode and Range

1) Mode: 3, Range: 4
2) Mode: 18, Range: 18
3) Mode: 8, Range: 14
4) Mode: 24, Range: 14
5) Mode: 27, Range: 21

6) Mode: 8, Range: 15
7) Mode: 13, Range: 13
8) Mode: 5, Range: 15
9) Mode: 7, Range: 14
10) Mode: 15, Range: 15

11) Mode: 6, Range: 40
12) Mode: 8, Range: 30
13) Mode: 12, Range: 45
14) Mode: 15, Range: 8
15) 10

Stem–And–Leaf Plot

1)

Stem	leaf
2	0 1 4 6 7 9
3	0 6 7
5	1 3 5

2)

Stem	leaf
1	1 4 8 9
5	9 9
6	1 5 5 6 8 9

3)

Stem	leaf
5	4 5 9
6	3 6 8
11	2 9
12	1 3 5 8

4)

Stem	leaf
3	2 6 9
5	1 6 6 6
8	4 5 9
10	0 6 7

5)

Stem	leaf
1	1 3 6 7 9
3	0 3 5 5
8	1 6 7 9

6)

Stem	leaf
2	2 1 5 9
6	0 2 4 7
9	2 3 5 8

Pie Graph

1) $5,600
2) $784

3) $1,288
4) $2,016

5) $\frac{1}{2}$

TSI Subject Test Mathematics

Probability Problems

1) $\frac{3}{5}$

2) $\frac{1}{4}, \frac{2}{3}$

3) $\frac{1}{4}$

4) $\frac{1}{13}$

5) $\frac{7}{15}$

6) $\frac{1}{2}$

7) $\frac{1}{2}$

8) $\frac{1}{8}$

9) 0

Factorials

1) 25

2) 122

3) 4

4) 114

5) 724

6) 9

7) 64

8) 324

9) 26

10) 90

11) 30

12) 6

13) 210

14) $n(n-1)(n-2)$

15) $(n+1)(n+2)$

16) 32

17) $5(n+2)$

18) 19.25

19) 26

20) 3

21) 16

22) $\frac{11}{13}$

23) 3,600

24) 15

25) $\frac{24}{5n(5n-1)}$

26) $\frac{n}{(n+2)}$

27) $\frac{1}{n(n-1)(n+2)}$

Combinations and Permutations

1) 720

2) 240

3) 72

4) 126

5) 5,040

6) 24

7) 12

8) 4,920

9) 120

10) 120

11) 24

12) 120

13) 720

14) 5,040

15) 120

16) 42

Chapter 13 :

Geometry and Solid Figures

Topics that you will practice in this chapter:

- ✓ Angles
- ✓ Pythagorean Relationship
- ✓ Triangles
- ✓ Polygons
- ✓ Trapezoids
- ✓ Circles
- ✓ Cubes
- ✓ Rectangular Prism
- ✓ Cylinder
- ✓ Pyramids and Cone

Geometry is the archetype of the beauty of the world.

Johannes Kepler

Angles

✍ **What is the value of x in the following figures?**

1)

2)

3)

4)

5)

6)
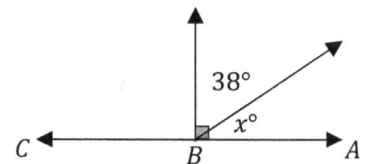

✍ **Calculate.**

7) Two supplement angles have equal measures. What is the measure of each angle? _____

8) The measure of an angle is seven fifth the measure of its supplement. What is the measure of the angle? _____

9) Two angles are complementary and the measure of one angle is 24 less than the other. What is the measure of the smaller angle? _____

10) Two angles are complementary. The measure of one angle is one fifth the measure of the other. What is the measure of the bigger angle? _____

11) Two supplementary angles are given. The measure of one angle is 40° less than the measure of the other. What does the smaller angle measure? _____

Pythagorean Theorem

✍ **Do the following lengths form a right triangle?**

1)

2)

3)

4)

5)

6)

7)

8)

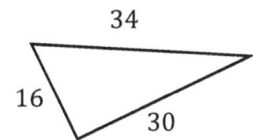

✍ **Find the missing side?**

9)

10)

11)

12)

13)

14)

15)

16)

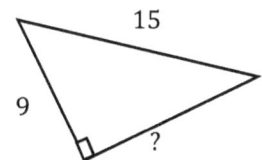

Triangles

✎ **Find the measure of the unknown angle in each triangle.**

1)

38°
81°
?°

2)

45°
87°
?°

3)

40°
85°
?°

4)

56°
72°
?°

5)

40°
95°
?°

6)

30°
110°
?°

7)

?°
35°
100°

8)
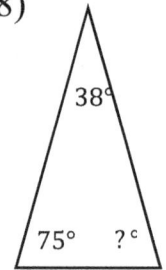
38°
75°
?°

✎ **Find area of each triangle.**

9)

15
9
12

10)

26
24
10

11)
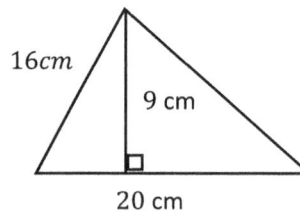
16cm
9 cm
20 cm

12)
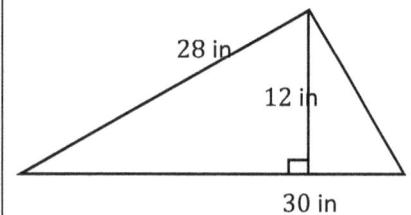
28 in
12 in
30 in

Polygons

✏️ **Find the perimeter of each shape.**

1)

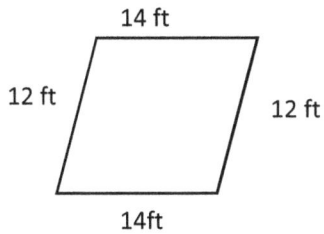

14 ft
12 ft
12 ft
14ft

2)

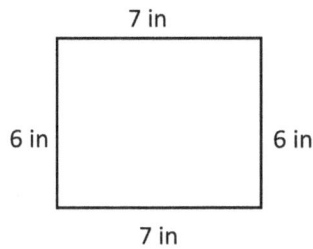

7 in
6 in
6 in
7 in

3)

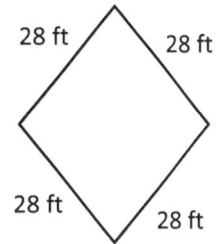

28 ft   28 ft
28 ft   28 ft

4) Square

5 cm

5) Regular hexagon

9 m

6)

5.3 cm
7.2 cm
4 cm
7.2 cm
5.3 cm

7) Parallelogram

7 in
12 in

8) Square

6 m

✏️ **Find the area of each shape.**

9) Parallelogram

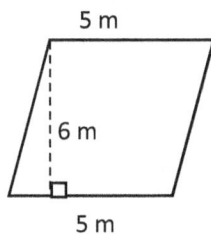

5 m
6 m
5 m

10) Rectangle

25 in
12 in

11) Rectangle

16 km
10 km

12) Square

7 in

Trapezoids

✎ **Find the area of each trapezoid.**

1)

2)

3)

4)

5)

6)

7)

8)

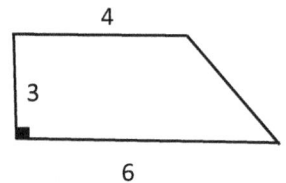

✎ **Calculate.**

1) A trapezoid has an area of 45 cm² and its height is 5 cm and one base is 5 cm. What is the other base length? _____

2) If a trapezoid has an area of 99 ft² and the lengths of the bases are 8 ft and 10 ft, find the height? _____

3) If a trapezoid has an area of 126 m² and its height is 14 m and one base is 6 m, find the other base length? _____

4) The area of a trapezoid is 440 ft² and its height is 22 ft. If one base of the trapezoid is 15 ft, what is the other base length?

_____

Circles

✍ **Find the area of each circle.** ($\pi = 3.14$)

1)	2)	3)	4)	5)	6)
2.5 in	5cm	9 ft	2 m	11 cm	10 miles

7)	8)	9)	10)	11)	12)
13 in	3 ft	4 m	12 cm	7 miles	8 ft

✍ **Complete the table below.** ($\pi = 3.14$)

Circle No.	Radius	Diameter	Circumference	Area
1	1 in	2 in	6.28 in	3.14 in^2
2		10 m		
3				28.26 ft^2
4			47.1 mi	
5		11 km		
6	7 cm			
7		12 ft		
8				314 m^2
9			56.52 in	
10	4.5 ft			

Cubes

✎ **Find the volume of each cube.**

1)	2)	3)	4)	5)	6)
	2 cm	6 ft	11 m	13 in	7 miles

7)	8	9)	10)	11)	12)
1.2 km	9 cm	2.1 ft	12 mm	0.2 in	0.1 km

✎ **Find the surface area of each cube.**

13)	14)	15)	16)	17)	18)
	7 m	5 ft	4.5 mm	1.1 km	11 cm

Rectangular Prism

✎ **Find the volume of each Rectangular Prism.**

1)

11 m
4 m
3 m

2)

16 in
10 in
5 in

3)

15 m
5 m
8 m

4)

2 cm
7 cm
8 cm

5)

3 ft
12 ft
8 ft

6)

6 m
10 m
7 m

✎ **Find the surface area of each Rectangular Prism.**

7)

8 cm
5 cm
4 cm

8)

6 ft
12 ft
3 ft

9)

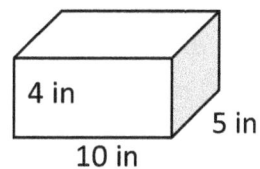

4 in
10 in
5 in

10)

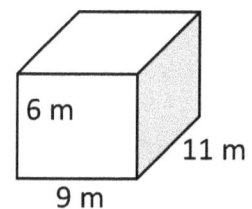

6 m
11 m
9 m

Cylinder

✎ **Find the volume of each Cylinder. Round your answer to the nearest tenth. ($\pi = 3.14$)**

1)

16 m
5m

2)

15.5 cm
4.2 cm

3)

12 cm
21 cm

4)

$\frac{5}{8}$m
$\frac{9}{10}$m

5)

30 m
2.5 m

6)
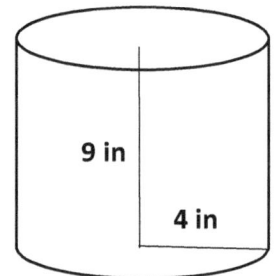
9 in
4 in

✎ **Find the surface area of each Cylinder. ($\pi = 3.14$)**

7)

7 m
3 m

8)

10 cm
6 cm

9)

1 cm
5 cm

10)
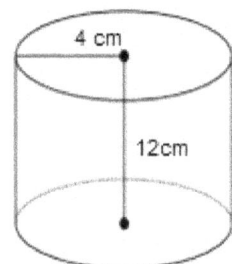
4 cm
12cm

Pyramids and Cone

✎ **Find the volume of each Pyramid and Cone.** ($\pi = 3.14$)

1)

2)

3)

4)

5)

6)

✎ **Find the surface area of each Pyramid and Cone.** ($\pi = 3.14$)

7)

8)

9)

10)

Answers of Worksheets

Angles

1) 16°   4) 34°   7) 90°   10) 75°
2) 96°   5) 70°   8) 75°   11) 70°
3) 59°   6) 52°   9) 33°

Pythagorean Relationship

1) No   5) Yes   9) 13   13) 15
2) Yes   6) No   10) 20   14) 30
3) No   7) Yes   11) 17   15) 36
4) Yes   8) Yes   12) 10   16) 12

Triangles

1) 60°   5) 45°   9) 54 *square unites*
2) 48°   6) 40°   10) 120 *square unites*
3) 55°   7) 45°   11) 90 *square unites*
4) 52°   8) 67°   12) 180 *square unites*

Polygons

1) 52 ft 5) 54 m 9) 30 m^2
2) 26 in 6) 25 cm 10) 300 in^2
3) 112 ft 7) 38 in 11) 160 km^2
4) 20 cm 8) 24 m 12) 49 in^2

Trapezoids

1) 50 cm^2 4) 60 cm^2 7) 36
2) 105 m^2 5) 80   8) 15
3) 39 ft^2 6) 24

Calculate

1) 13 cm 2) 11 ft 3) 12 m 4) 25 ft

Circles

1) 19.63 in^2 5) 379.94 cm^2 9) 12.56 m^2
2) 78.5 cm^2 6) 314 $miles^2$ 10) 113.04 cm^2
3) 254.34 ft^2 7) 132.67 in^2 11) 38.47 $miles^2$
4) 12.56 m^2 8) 7.07 ft^2 12) 50.24 ft^2

Circle No.	Radius	Diameter	Circumference	Area
1	1 in	2 in	6.28 in	3.14 in^2
2	5 m	10 m	31.4 m	78.5 m^2
3	3 ft	6 ft	18.84 ft	28.26 ft^2
4	7.5 miles	15 mi	47.1 mi	176.63 mi^2
5	5.5 km	11 km	34.54 km	94.99 km^2
6	7 cm	14 cm	43.96 cm	153.86 cm^2
7	6 ft	12 ft	37.68 feet	113.04 ft^2
8	10 m	20 m	62.8 m	314 m^2
9	9 in	18 in	56.52 in	254.34 in^2
10	4.5 ft	9 ft	28.26 ft	63.585 ft^2

Cubes

1) 12
2) 8 cm^3
3) 216 ft^3
4) 1,331 m^3
5) 2,197 in^3
6) 343 $miles^3$
7) 1.728 km^3
8) 729 cm^3
9) 9.261 ft^3
10) 1,728 mm^3
11) 0.008 in^3
12) 0.001 km^3
13) 27
14) 294 m^2
15) 150 ft^2
16) 121.5 mm^2
17) 7.26 km^2
18) 726 cm^2

Rectangular Prism

1) 132 m^3
2) 800 in^3
3) 600 m^3
4) 112 cm^3
5) 288 ft^3
6) 420 m^3
7) 184 cm^2
8) 252 ft^2
9) 220 in^2
10) 438 m^2

Cylinder

1) 1,004.8 m^3
2) 214.6 cm^3
3) 9,495.4 cm^3
4) 1.1 m^3
5) 588.8 m^3
6) 452.2 in^3
7) 188.4 m^2
8) 602.9 cm^2
9) 37.7 cm^2
10) 401.9 m^2

Pyramids and Cone

1) 1,600 yd^3
2) 1,050 yd^3
3) 1,617 in^3
4) 392.5 m^3
5) 3,014.4 m^3
6) 366.33 cm^3
7) 1,440 yd^2
8) 1,536 m^2
9) 678.24 in^2
10) 1,205.76 cm^2

Chapter 14 :
TSI Math Test Review

The Texas Success Initiative Assessment, is known as the TSI, is a test to determine the appropriate level of college course work for an incoming student. In essence, it is a broad and quick assessment of students' academic abilities.

The TSI test consists of three separate exams:

- ✓ Mathematics
- ✓ Reading
- ✓ Writing

The mathematics portion of the TSI test contains 20 questions. The test covers data analysis, geometry, and algebra on both intermediate and basic levels.

Students are not allowed to use calculator when taking a TSI assessment. A pop-up calculator is embedded in the test for some questions.

In this section, there are two complete TSI Mathematics Tests. Take these tests to see what score you'll be able to receive on a real TSI test.

The hardest arithmetic to master is that which enables us to count our blessings.
Eric Hoffer

Time to Test

Time to refine your skill with a practice examination.

Take a practice TSI Math Test to simulate the test day experience. After you've finished, score your test using the answer key.

Before You Start

- You'll need a pencil, a calculator and a timer to take the test.

- For each question, there are four possible answers. Choose which one is best.

- Use the answer sheet provided to record your answers.

- After you've finished the test, review the answer key to see where you went wrong.

Good Luck!

TSI Math Practice Test Answer Sheets

Remove (or photocopy) these answer sheets and use them to complete the practice tests.

TSI Mathematics Practice Test

Section 1		Section 2	
1	Ⓐ Ⓑ Ⓒ Ⓓ Ⓔ	1	Ⓐ Ⓑ Ⓒ Ⓓ Ⓔ
2	Ⓐ Ⓑ Ⓒ Ⓓ Ⓔ	2	Ⓐ Ⓑ Ⓒ Ⓓ Ⓔ
3	Ⓐ Ⓑ Ⓒ Ⓓ Ⓔ	3	Ⓐ Ⓑ Ⓒ Ⓓ Ⓔ
4	Ⓐ Ⓑ Ⓒ Ⓓ Ⓔ	4	Ⓐ Ⓑ Ⓒ Ⓓ Ⓔ
5	Ⓐ Ⓑ Ⓒ Ⓓ Ⓔ	5	Ⓐ Ⓑ Ⓒ Ⓓ Ⓔ
6	Ⓐ Ⓑ Ⓒ Ⓓ Ⓔ	6	Ⓐ Ⓑ Ⓒ Ⓓ Ⓔ
7	Ⓐ Ⓑ Ⓒ Ⓓ Ⓔ	7	Ⓐ Ⓑ Ⓒ Ⓓ Ⓔ
8	Ⓐ Ⓑ Ⓒ Ⓓ Ⓔ	8	Ⓐ Ⓑ Ⓒ Ⓓ Ⓔ
9	Ⓐ Ⓑ Ⓒ Ⓓ Ⓔ	9	Ⓐ Ⓑ Ⓒ Ⓓ Ⓔ
10	Ⓐ Ⓑ Ⓒ Ⓓ Ⓔ	10	Ⓐ Ⓑ Ⓒ Ⓓ Ⓔ
11	Ⓐ Ⓑ Ⓒ Ⓓ Ⓔ	11	Ⓐ Ⓑ Ⓒ Ⓓ Ⓔ
12	Ⓐ Ⓑ Ⓒ Ⓓ Ⓔ	12	Ⓐ Ⓑ Ⓒ Ⓓ Ⓔ
13	Ⓐ Ⓑ Ⓒ Ⓓ Ⓔ	13	Ⓐ Ⓑ Ⓒ Ⓓ Ⓔ
14	Ⓐ Ⓑ Ⓒ Ⓓ Ⓔ	14	Ⓐ Ⓑ Ⓒ Ⓓ Ⓔ
15	Ⓐ Ⓑ Ⓒ Ⓓ Ⓔ	15	Ⓐ Ⓑ Ⓒ Ⓓ Ⓔ
16	Ⓐ Ⓑ Ⓒ Ⓓ Ⓔ	16	Ⓐ Ⓑ Ⓒ Ⓓ Ⓔ
17	Ⓐ Ⓑ Ⓒ Ⓓ Ⓔ	17	Ⓐ Ⓑ Ⓒ Ⓓ Ⓔ
18	Ⓐ Ⓑ Ⓒ Ⓓ Ⓔ	18	Ⓐ Ⓑ Ⓒ Ⓓ Ⓔ
19	Ⓐ Ⓑ Ⓒ Ⓓ Ⓔ	19	Ⓐ Ⓑ Ⓒ Ⓓ Ⓔ
20	Ⓐ Ⓑ Ⓒ Ⓓ Ⓔ	20	Ⓐ Ⓑ Ⓒ Ⓓ Ⓔ

TSI Practice Test 1

Section 1

Arithmetic

❖ **20 Questions.**

❖ **Total time for this test: No time limit**.

❖ **You may NOT use a calculator on this Section.**

(On a real Accuplacer test, there is an onscreen calculator to use on some

questions).

Administered *Month Year*

TSI Subject Test Mathematics

1) Which of the following is closest to 8.12?

    A. 7

    B. 9

    C. 8

    D. 9.1

2) Which of the following is greater than $\frac{14}{5}$?

    A. $\frac{1}{2}$

    B. $\frac{9}{2}$

    C. $\frac{5}{2}$

    D. 1

3) What is the value of 2.85×8.2?

    A. 23.37

    B. 2.337

    C. 233.7

    D. 33.37

4) Which of the following inequalities is true?

    A. $\frac{16}{26} < \frac{7}{13}$

    B. $\frac{5}{9} < \frac{1}{3}$

    C. $\frac{5}{6} < \frac{23}{42}$

    D. $\frac{11}{24} < \frac{7}{8}$

5) 45% of what number is equal to 54?

    A. 1.2

    B. 12

    C. 120

    D. 100

6) The price of a sofa is decreased by 38% to $279. What was its original price?

    A. $106.02

    B. $102.06

    C. $450

    D. $480

7) If 65% of a class are girls, and 20% of girls play tennis, what percent of the class play tennis?

A. 13%   C. 10%

B. 14%   D. 87%

8) If 84% of A is 14% of B, then B is what percent of A?

A. 6%   C. 600%

B. 60%   D. 0.6%

9) A taxi driver earns $9 per 1-hour work. If he works 10 hours a day and in 1 hour, he uses 1.5-liters petrol with price $3 for 1-liter. How much money does he earn in one day?

A. $43   C. $54

B. $42   D. $45

10) The price of a car was $54,000 in 2014, $40,500 in 2016 and $30,375 in 2017. What is the rate of depreciation of the price of car per year?

A.15%   C.30%

B. 25%   D. 20%

11) Which of the following expressions has the same value as $\frac{24}{7} \times \frac{14}{5}$?

A. $\frac{8 \times 4}{5}$ C. $\frac{11 \times 4}{5}$

B. $\frac{12 \times 2}{5}$ D. $\frac{24 \times 2}{5}$

12) A bank is offering 2.5% simple interest on a savings account. If you deposit $35,000, how much interest will you earn in four years?

A. $250

C. $3,500

B. $350

D. $2,500

13) Which of the following lists shows the fractions in order from least to greatest?

$$\frac{4}{19}, \frac{7}{12}, \frac{8}{9}, \frac{5}{16}$$

A. $\frac{5}{16}, \frac{8}{9}, \frac{7}{12}, \frac{4}{19}$

C. $\frac{4}{19}, \frac{5}{16}, \frac{7}{12}, \frac{8}{9}$

B. $\frac{8}{9}, \frac{7}{12}, \frac{4}{19}, \frac{5}{16}$

D. $\frac{4}{19}, \frac{7}{12}, \frac{5}{16}, \frac{8}{9}$

14) Sophia purchased a sofa for $472.50. The sofa is regularly priced at $630. What was the percent discount Sophia received on the sofa?

A. 20%

C. 30%

B. 25%

D. 15%

15) A rope weighs 500 grams per meter of length. What is the weight in kilograms of 21.4 meters of this rope? (1 kilograms = 1,000 grams)

A. 0.107

C. 10.7

B. 1.07

D. 107

16) $\frac{19}{6} - \frac{14}{3} = ?$

A. −0.75

C. 1.5

B. 0.75

D. −1.5

17) When 61 is divided by 7, the remainder is the same as when 41 is divided by

    A. 11   C. 9

    B. 10   D. 8

18) 24 is What percent of 60?

    A. 60%   C. 40%

    B. 80%   D. 140%

19) How long does a 620–miles trip take moving at 50 miles per hour (mph)?

    A. 12 hours   C. 12 hours and 24 minutes

    B. 12 hours and 4 minutes   D. 12 hours and 40 minutes

20) What is the value of $7.51 + 0.234 + 0.6367$?

    A. 8.8307   C. 8.3708

    B. 8.3807   D. 8.3087

STOP

This is the End of This Section.

You may check your work on this Section if you still have time.

TSI Practice Test 1

Section 2

Quantitative Reasoning, Algebra, and Statistics

❖ **20 Questions.**

❖ **Total time for this test: No time limit**.

❖ **You may NOT use a calculator on this Section.**

(On a real Accuplacer test, there is an onscreen calculator to use on some questions).

Administered *Month Year*

1) Four seventh of 35 is equal to $\frac{5}{9}$ of what number?

    A. 28   C. 35

    B. 36   D. 38

2) A boat sails 72 miles south and then 30 miles east. How far is the boat from its start point?

    A. 87 miles   C. 94 miles

    B. 78 miles   D. 81 miles

3) The ratio of boys and girls in a class is 3:7. If there are 90 students in the class, how many more boys should be enrolled to make the ratio 1:1?

    A. 17   C. 36

    B. 34   D. 27

4) The score of Emma was quarter as that of Ava and the score of Mia was twice that of Ava. If the score of Mia was 80, what is the score of Emma?

    A. 15   C. 5

    B. 10   D. 40

5) An angle is equal to one fifth of its supplement. What is the measure of that angle?

    A. 30   C. 15

    B. 36   D. 20

6) When a number is subtracted from 70 and the difference is divided by that number, the result is 9. What is the value of the number?

   A. 8   C. 7

   B. 9   D. 5

7) John traveled 150 km in 5 hours and Alice traveled 280 km in 4 hours. What is the ratio of the average speed of John to average speed of Alice?

   A. 7: 3   C. 3: 7

   B. 5: 8   D. 9: 5

8) A taxi driver earns \$5 per 1-hour work. If he works 12 hours a day and in 1 hour, he uses 2.5-liters petrol with price \$1.2 for 1-liter. How much money does he earn in one day?

   A. \$96   C. \$24

   B. \$36   D. \$26

9) If the area of trapezoid is 162, what is the perimeter of the trapezoid?

   A. 25

   B. 52

   C. 52.50

   D. 25.50

10) Right triangle ABC has two legs of lengths 15 cm (AB) and 36 cm (AC). What is the length of the third side (BC)?

    A. 45 cm   C. 35 cm

    B. 42 cm   D. 39 cm

11) The area of a circle is less than 169π. Which of the following can be the circumference of the circle?

    A. 26π C. 30π

    B. 27π D. 34π

12) The width of a box is one sixth of its length. The height of the box is one fourth of its width. If the length of the box is 120 cm, what is the volume of the box?

    A. 9,600 cm^3 C. 15,000 cm^3

    B. 2,400 cm^3 D. 12,000 cm^3

13) How many possible outfit combinations come from six shirts, seven slacks, and five times?

    A. 42   C. 120

    B. 105   D. 210

14) In the xy-plane, the point (2, 2) and (4, 6) are on the line A. Which of the following points could also be on the line A? (Select one or more answer choices)

    A. $(-1, 5)$ C. $(-2, -3)$

    B. $(2, 6)$ D. $(-5, -12)$

15) A bag contains 15 balls: five green, two black, four blue, a brown, a red and two white. If 8 balls are removed from the bag at random, what is the probability that a brown ball has been removed?

A. $\frac{2}{15}$

C. $\frac{8}{15}$

B. $\frac{1}{8}$

D. $\frac{1}{15}$

16) The average of five consecutive numbers is 84. What is the smallest number?

A. 86

C. 82

B. 83

D. 84

17) The average weight of 18 girls in a class is 45 kg and the average weight of 32 boys in the same class is 55 kg. What is the average weight of all the 50 students in that class?

A. 52

C. 52.80

B 51.40

D. 51.80

18) What is the median of these numbers? 10, 24, 12, 18, 37, 29, 23

A.18

C. 12

B. 23

D. 29

19) The surface area of a cylinder is $160\pi\ cm^2$. If its height is 2 cm, what is the radius of the cylinder?

A. 10 cm

C. 6 cm

B. 12 cm

D. 8 cm

20) In 1999, the average worker's income increased $5,000 per year starting from

$36,000 annual salary. Which equation represents income greater than average?

(I = income, x = number of years after 1999)

A. $I > 5,000\, x + 36,000$

B. $I > -\, 5,000\, x + 36,000$

C. $I < -5,000\, x + 36,000$

D. $I < 5,000\, x - 36,000$

STOP

This is the End of This Section.

You may check your work on this Section if you still have time.

TSI Practice Test 2

Section 1

Arithmetic

❖ **20 Questions.**

❖ **Total time for this test: No time limit**.

❖ **You may NOT use a calculator on this Section.**

(On a real Accuplacer test, there is an onscreen calculator to use on some questions).

Administered *Month Year*

1) What is $53,126 + 7,492$?

    A. 60,618   C. 60,508

    B. 60,816   D. 60,718

2) Bob deposits 15% of $420 into a savings account, what is the amount of his deposit?

    A. $75   C. $36

    B. $57   D. $63

3) What is 4 percent of 450?

    A. 28   C. 18

    B. 10   D. 20

4) In two successive years, the population of a town is increased by 15% and 20%. What percent of the population is increased after two years?

    A. 64%   C. 36%

    B. 62%   D. 38%

5) For what price is 30 percent off the same as $90 off?

    A. $300   C. $150

    B. $450   D. $250

6) What is the remainder when 530 is divided by 7?

    A. 9   C. 5

    B. 6   D. 7

7) Last week 15,000 fans attended a football match. This week four times as many bought tickets, but one eighth of them cancelled their tickets. How many are attending this week?

    A. 54,375　　　　　　　　　　　　C. 52,500

    B. 52,000　　　　　　　　　　　　D. 45,500

8) Which of the following shows the numbers in descending order?

$$\frac{1}{5}, 0.83, 68\%, \frac{5}{14}$$

    A. $68\%, 0.83, \frac{1}{5}, \frac{5}{14}$　　　　　　C. $0.83, 68\%, \frac{1}{5}, \frac{5}{14}$

    B. $68\%, 0.83, \frac{5}{14}, \frac{1}{5}$　　　　　　D. $\frac{1}{5}, \frac{5}{14}, 68\%, 0.83$

9) If 160% of a number is 128, then what is the 45% of that number?

    A. 36　　　　　　　　　　　　　C. 63

    B. 22.5　　　　　　　　　　　　D. 25.5

10) Which of the following decimal numbers is less than $\frac{36}{32}$?

    A. 1.152　　　　　　　　　　　　C. 1.125

    B. 1.065　　　　　　　　　　　　D. 1.250

11) Mr. Jones saves $1,800 out of his monthly family income of $46,800. What fractional part of his income does he save?

    A. $\frac{1}{26}$　　　　　　　　　　　　C. $\frac{5}{24}$

    B. $\frac{1}{13}$　　　　　　　　　　　　D. $\frac{3}{26}$

12) 16% of what number is equal to 192?

    A. 30.72   C1,400

    B. 161.28   D. 1,200

13) 45 students took an exam and 36 of them failed. What percent of the students passed the exam?

    A. 35%   C. 65%

    B. 80%   D. 20%

14) A bank is offering 2.8% simple interest on a savings account. If you deposit $22,000, how much interest will you earn in seven years?

    A. $3,696   C. $4,200

    B. $4,312   D. $4,230

$$\frac{9}{20}, 0.453\%, 0.0453$$

15) Which of the following correctly orders the values above from least to greatest?

    A. $0.453\%, \frac{9}{20}, 0.0453$ C. $\frac{9}{20}, 0.453\%, 0.0453$

    B. $0.0453, 0.453\% \frac{9}{20}$ D. $\frac{9}{20}, 0.0453, 0.453\%$

16) Which of the following is equivalent to $\frac{2}{5}$?

    A. 0.04   C. 0.40

    B. 0.45   D. 4.0

17) If a gas tank can hold 32 gallons, how many gallons does it contain when it is $\frac{3}{4}$ full?

A. 26   C. 22.5

B. 18   D. 24

18) What is 0.2143 rounded to the nearest hundredth?

A. 0.214   C. 0.21

B. 0.215   D. 0.22

19) Jack earns $840 for his first 35 hours of work in a week and is then paid 1.5 times his regular hourly rate for any additional hours. This week, Jack needs $2,100 to pay his rent, bills and other expenses. How many hours must he work to make enough money in this week?

A. 24   C. 56

B. 35   D. 59

20) $2\frac{3}{4} - \frac{5}{8} = ?$

A. 1.152   C. 0.2125

B. 2.125   D. 2.152

STOP

This is the End of This Section.

You may check your work on this Section if you still have time.

TSI Practice Test 2

Section 2

Quantitative Reasoning, Algebra, and Statistics

❖ **20 Questions.**

❖ **Total time for this test: No time limit**.

❖ **You may NOT use a calculator on this Section.**

(On a real Accuplacer test, there is an onscreen calculator to use on some questions).

Administered *Month Year*

1) Simplify $10x^4y^4(2x^2y^3)^3 =$

   A. $40x^9y^{10}$ C. $80\ x^{10}y^{10}$

   B. $40x^9y^{13}$ D. $80\ x^{10}y^{13}$

2) The perimeter of a rectangular yard is 300 meters. What is its length if its width is five time its length?

   A. 25 meters   C. 30 meters

   B. 15 meters   D. 60 meters

3) What is the equivalent temperature of 59°F in Celsius?

   $C = \frac{5}{9}(F - 32)$

   A. 18   C. 20

   B. 15   D. 22

4) Right triangle ABC has two legs of lengths 10cm (AB) and 24cm (AC). What is the length of the third side (BC)?

   A. 18 cm   C. 37 cm

   B. 28 cm   D. 26 cm

5) Which of the following points lies on the line $2x + 3y = 7$?

   A. $(2, 1)$ C. $(-2, 1)$

   B. $(5, 1)$ D. $(3, -2)$

6) What is the area of a square whose diagonal is 8?

    A. 64   C. 42

    B. 32   D. 16

7) Two dice are thrown simultaneously, what is the probability of getting a sum of 4 or 6?

    A. $\frac{7}{36}$ C. $\frac{5}{9}$

    B. $\frac{2}{9}$ D. $\frac{3}{18}$

8) The mean of 40 test scores was calculated as 55. But it turned out that one of the scores was misread as 77 but it was 37. What is the correct mean of the test scores?

    A. 52   C. 54

    B. 45   D. 56

9) In a stadium the ratio of home fans to visiting fans in a crowd is 4:9. Which of the following could be the total number of fans in the stadium?

    A. 11,219   C. 24,729

    B. 13,453   D. 56,840

10) The perimeter of the trapezoid below is 60 cm. What is its area?

    A. 214 cm²

    B. 204 cm²

    C. 221 cm²

    D. 187 cm²

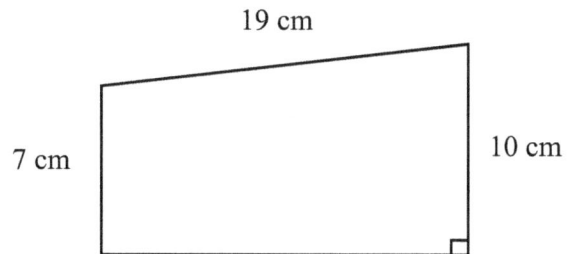

11) A swimming pool holds 1,800 cubic feet of water. The swimming pool is 25 feet long and 6 feet wide. How deep is the swimming pool?

    A. 8   C. 14

    B. 15   D. 12

12) Anita's trick–or–treat bag contains 9 pieces of chocolate, 14 suckers, 26 pieces of gum, 15 pieces of licorice. If she randomly pulls a piece of candy from her bag, what is the probability of her pulling out a piece of sucker?

    A. $\dfrac{7}{32}$ C. $\dfrac{5}{32}$

    B. $\dfrac{14}{65}$ D. $\dfrac{15}{64}$

13) The average of 9 numbers is 20. The average of 6 of those numbers is 17. What is the average of the other three numbers?

    A. 39   C. 22

    B. 32   D. 26

14) A card is drawn at random from a standard 52–card deck, what is the probability that the card is of hearts and diamonds? (The deck includes 13 of each suit clubs, diamonds, hearts, and spades)

A. $\frac{1}{52}$

C. $\frac{1}{13}$

B. $\frac{1}{2}$

D. $\frac{2}{13}$

15) A football team had $53,000 to spend on supplies. The team spent $45,000 on new balls. New sport shoes cost $280 each. Which of the following inequalities represent the number of new shoes the team can purchase?

A. $280x + 45,000 \leq 53,000$

C. $45,000x + 280 \leq 53,000$

B. $280x + 45,000 \geq 53,000$

D. $45,000x + 280 \geq 53,000$

16) The average of five numbers is 30. If a sixth number that is greater than 60 is added, then, which of the following could be the new average? (Select one or more answer choices)

A. 30

C. 34

B. 32

D. 36

17) The length of a rectangle is 3 meters less than 8 times its width. The perimeter of the rectangle is 102 meters. What is the area of the rectangle in meters?

A. 39

C. 210

B. 270

D. 240

18) The ratio of boys and girls in a class is 4:11. If there are 105 students in the class, how many more boys should be enrolled to make the ratio 1:1?

A. 28

C. 49

B. 77

D. 46

19) What is the value of x in the following equation?

$$\frac{4}{7}x + \frac{1}{6} = \frac{1}{2}$$

A. 7

C. $\frac{5}{12}$

B. $\frac{5}{6}$

D. $\frac{7}{12}$

20) What is the surface area of the cylinder below?

A. $280\ \pi\ in^2$

B. $160\ \pi\ in^2$

C. $480\ \pi\ in^2$

D. $240\ \pi\ in^2$

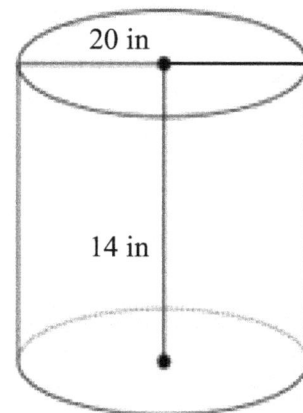

20 in

14 in

STOP

This is the End of This Section.

You may check your work on this Section if you still have time.

Chapter 15 :
Answers and Explanations

Answer Key

❋ Now, it's time to review your results to see where you went wrong and what areas you need to improve!

TSI Math Practice Test

	Practice Test 1				Practice Test 2		
	Section 1		**Section 2**		**Section 1**		**Section 2**
1	C	1	B	1	A	1	D
2	B	2	B	2	D	2	A
3	A	3	C	3	C	3	B
4	D	4	B	4	D	4	D
5	C	5	A	5	A	5	A
6	C	6	C	6	C	6	B
7	A	7	C	7	C	7	B
8	C	8	C	8	D	8	C
9	D	9	B	9	A	9	A
10	B	10	D	10	B	10	B
11	D	11	A	11	A	11	D
12	C	12	D	12	D	12	A
13	C	13	D	13	D	13	D
14	B	14	D	14	B	14	B
15	C	15	D	15	D	15	A
16	D	16	C	16	C	16	D
17	C	17	B	17	D	17	B
18	C	18	B	18	C	18	C
19	C	19	D	19	D	19	D
20	B	20	A	20	B	20	C

Practice Tests 1

Section 1

1) Answer: C.

The closest to 8.12 is 8 in the options provided.

2) Answer: B.

$\frac{14}{5} = 2.8$, the only option that is greater than 2.8 is $\frac{9}{2} = 4.5, 4.5 > 2.8$

3) Answer: A.

First multiply the tenths place of 0.2 by 2.85. The result is 0.57. Next, multiply 8 by 2.85 which results in 22.8. The sum of these two numbers is: $0.57 + 22.8 = 23.37$

4) Answer: D.

To compare fractions, find a common denominator. When two fractions have common denominators, the fraction with the larger numerator is the larger number. Choice A is incorrect because $\frac{16}{26}$ is not less than $\frac{7}{13}$. Write both fractions with common denominator and compare the numerators. $\frac{7}{13} = \frac{14}{26}$.

The fraction $\frac{16}{26}$ is greater than $\frac{7}{13}$.

Choice B and C are not correct. Shown written with a common denominator, the comparisons $\frac{5}{9} < \frac{1}{3}$ and $\frac{5}{6} < \frac{23}{42}$ are not correct. Shown written with a common denominator, the comparison $\frac{11}{24} < \frac{7}{8}$ is correct because $\frac{7}{8}$ or $\frac{21}{24}$ is greater than $\frac{11}{24}$.

5) Answer: C.

Dividing 54 by 45%, which is equivalent to 0.45, gives 120. Therefore, 45% of 120 is 54.

6) Answer: C.

Let x be the original price. If the price of the sofa is decreased by 38% to $279, then:

$62\%\ of\ x = 279 \Rightarrow 0.62x = 279 \Rightarrow x = 279 \div 0.62 = 450$

7) Answer: A.

The percent of girls playing tennis is: $65\% \times 20\% = 0.65 \times 0.20 = 0.13 = 13\%$

8) Answer: C.

Write the equation and solve for B: $0.84\ A = 0.14\ B$, divide both sides by 0.14, then:

$\frac{0.84}{0.14}A = B$, therefore: $B = 6A$, and B is 6 times of A or it's 600% of A.

9) Answer: D.

$\$9 \times 10 = \$90,$ Petrol use:$10 \times 1.5 = 15$ liters

Petrol cost: $15 \times \$3 = \$45.$ Money earned: $\$90 - \$45 = \$45$

10) Answer: B.

Use this formula: Percent of Change $\frac{\text{New Value} - \text{Old Value}}{\text{Old Value}} \times 100\%$

$\frac{40,500 - 54,000}{54,000} \times 100\% = -25\%$ and $\frac{30,375 - 40,500}{40,500} \times 100\% = -25\%$

11) Answer: D.

First simplify the multiplication: $\frac{24}{7} \times \frac{14}{5} = \frac{48}{5} = \frac{48}{5}$

Choice D is equal to $\frac{48}{5}$. $\frac{24 \times 2}{5} = \frac{48}{5}$

12) Answer: C.

Use simple interest formula:

$I = prt$ ($I = interest,\ p = principal, r = rate, t = time$)

$I = (35,000)(0.025)(4) = 3,500$

13) Answer: C.

Let's compare each fraction: $\frac{4}{19} < \frac{5}{16} < \frac{7}{12} < \frac{8}{9}$ Only choice C provides the right order.

14) Answer: B.

The question is this: 472.50 is what percent of 630? Use percent formula:

$\text{part} = \frac{\text{percent}}{100} \times \text{whole}$

$472.50 = \frac{\text{percent}}{100} \times 630 \Rightarrow 472.50 = \frac{\text{percent} \times 630}{100} \Rightarrow 47,250 = \text{percent} \times 630 \Rightarrow$

$\text{percent} = \frac{47,250}{630} = 75$

472.50 is 75% of 630. Therefore, the discount is:

$100\% - 75\% = 25\%$

15) Answer: C.

The weight of 21.4 meters of this rope is: $21.4 \times 500\ g = 10,700\ g$

1 kg = 1,000 g, therefore, $10,700\ g \div 1,000 = 10.7\ kg$

16) Answer: D.

$\frac{19}{6} - \frac{28}{6} = -\frac{9}{6} = -\frac{3}{2} = -1.5$

17) Answer: C.

61 divided by 7, the remainder is 5. 41 divided by 9, the remainder is also 5.

18) Answer: C.

Use percent formula: $\text{part} = \frac{\text{percent}}{100} \times \text{whole}$

$24 = \frac{\text{percent}}{100} \times 60 \Rightarrow 24 = \frac{\text{percent} \times 60}{100} \Rightarrow 24 = \frac{\text{percent} \times 6}{10}$, multiply both sides by 10.

$240 = \text{percent} \times 6$, divide both sides by 6. $40 = \text{percent}$

19) Answer: C.

Use distance formula: Distance = Rate × time $\Rightarrow 620 = 50 \times T$, divide both sides by 50. $\frac{620}{50} = T \Rightarrow T = 12.4$ hours.

Change hours to minutes for the decimal part. 0.4 hours = 0.4 × 60 = 24 minutes.

20) Answer: B.

To add decimal numbers, line them up and add from right.

$7.51 + 0.234 + 0.6367 = 8.3807$

Practice Tests 1

Section 2

1) Answer: B.

Let x be the number. Write the equation and solve for x.

$\frac{4}{7} \times 35 = \frac{5}{9} . x \Rightarrow \frac{4 \times 35}{7} = \frac{5x}{9}$, use cross multiplication to solve for x.

$9 \times 4 \times 35 = 5x \times 7 \Rightarrow 1,260 = 35x \Rightarrow x = 36$

2) Answer: B.

Use the information provided in the question to draw the shape.

Use Pythagorean Theorem: $a^2 + b^2 = c^2$

$72^2 + 30^2 = c^2 \Rightarrow 5,184 + 900 = c^2 \Rightarrow 6,084 = c^2 \Rightarrow c = 78$

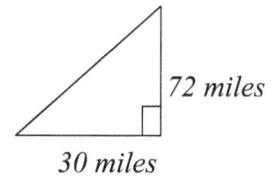

72 miles

30 miles

3) Answer: C.

The ratio of boy to girls is 3:7. Therefore, there are 3 boys out of 10 students. To find the answer, first divide the total number of students by 10, then multiply the result by 3.

$90 \div 10 = 9 \Rightarrow 9 \times 3 = 27$ There are 27 boys and $63(90 - 27)$ girls. So, 36 more boys should be enrolled to make the ratio 1:1

4) Answer: B.

If the score of Mia was 80, therefore the score of Ava is 40. Since, the score of Emma was quarter as that of Ava, therefore, the score of Emma is 10.

5) Answer: A.

The sum of supplement angles is 180. Let x be that angle. Therefore, $x + 5x = 180$

$6x = 180$, divide both sides by 6: $x = 30$

6) Answer: C.

Let x be the number. Write the equation and solve for x. $(70 - x) \div x = 9$

Multiply both sides by x. $(70 - x) = 9x$, then add x both sides. $70 = 10x$, now divide both sides by 10. $x = 7$

7) Answer: C.

The average speed of john is: $150 \div 5 = 30 \ km$

The average speed of Alice is: $280 \div 4 = 70 \ km$

Write the ratio and simplify. $30 : 70 \Rightarrow 3:7$

8) Answer: C.

$5 \times 12 = \$60$,   Petrol use: $12 \times 2.5 = 30$ liters

Petrol cost: $30 \times \$1.2 = \$36$

Money earned: $\$60 - \$36 = \$24$

9) Answer: B.

The area of the trapezoid is:

Area$= \frac{1}{2}h(b_1 + b_2) = \left(\frac{11+16}{2}\right) \times x = 162 \rightarrow 13.5x = 162 \rightarrow x = 12$

$y = \sqrt{5^2 + 12^2} = \sqrt{169} = 13$

Perimeter is: $13 + 5 + 11 + 12 + 11 = 52$

10) Answer: D.

Use Pythagorean Theorem: $a^2 + b^2 = c^2$

$15^2 + 36^2 = c^2 \Rightarrow 225 + 1{,}296 = c^2 \Rightarrow 1{,}521 = c^2 \Rightarrow c = 39$

11) Answer: A

Area of the circle is less than than $169\,\pi$. Use the formula of areas of circles.

Area $= \pi r^2 \Rightarrow 169\pi > \pi r^2 \Rightarrow 169 > r^2 \Rightarrow r < 13$

Radius of the circle is less than 13. Let's put 12 for the radius. Now, use the circumference formula:

Circumference $= 2\pi r = 2\pi\,(13) = 26\pi$

Since the radius of the circle is less than 13. Then, the circumference of the circle must be less than 26π. Only choice A is less than 26π.

12) Answer: D.

If the length of the box is 120, then the width of the box is one sixth of it, 20, and the height of the box is 5 (one fourth of the width). The volume of the box is:

V = (length) × (width) × (height) = lwh = (120) × (20) × (5) = 12,000 cm^3

13) Answer: D.

To find the number of possible outfit combinations, multiply number of options for each factor: $6 \times 7 \times 5 = 210$

14) Answer: D.

The equation of a line is in the form of $y = mx + b$, where m is the slope of the line and b is the $y - intercept$ of the line.

Two points $(2, 2)$ and $(4, 6)$ are online A. Therefore, the slope of the line A is:

slope of line A $= \frac{y_2 - y_1}{x_2 - x_1} = \frac{6-2}{4-2} = \frac{4}{2} = 2$

The slope of line A is 2. Thus, the formula of the line A is:

$y = mx + b = 2x + b$, choose a point and plug in the values of x and y in the equation to solve for b. Let's choose point $(2, 2)$. Then:$y = 2x + b \rightarrow 2 = 4 + b \rightarrow b = 2 - 4 = -2$

The equation of line A is: $y = 2x - 2$

Now, let's review the choices provided:

A. $(-1, 5)$ $y = 2x - 2 \rightarrow 5 = -2 - 2 = -4$ This is not true.

B. $(2, 6)$ $y = 2x - 2 \rightarrow 6 = 4 - 2 = 2$ This is not true.

C. $(-2, -3)$ $y = 2x - 2 \rightarrow -3 = -4 - 2 = -6$ This is not true.

D. $(-5, -12)$ $y = 2x - 2 \rightarrow -12 = -10 - 2 = -12$ This is true!

15) Answer: D.

If 12 balls are removed from the bag at random, there will be one ball in the bag.

The probability of choosing a brown ball is 1 out of 15. Therefore, the probability of not choosing a brown ball is 8 out of 15 and the probability of having not a brown ball after removing 8 balls is the same.

16) Answer: C.

Let x be the smallest number. Then, these are the numbers: $x, x + 1, x + 2, x + 3, x + 4,$

average $= \frac{\text{sum of terms}}{\text{number of terms}} \Rightarrow 84 = \frac{x+(x+1)+(x+2)+(x+3)+(x+4)}{5} \Rightarrow 84 = \frac{5x+10}{5} \Rightarrow 420 = 5x + 10$

$\Rightarrow 410 = 5x \Rightarrow x = 82$

17) Answer: B.

average $= \frac{\text{sum of terms}}{\text{number of terms}}$

The sum of the weight of all girls is: $18 \times 45 = 810 \ kg$

The sum of the weight of all boys is: $32 \times 55 = 1,760 \ kg$

The sum of the weight of all students is: $810 + 1,760 = 2,570 \ kg$

average $= \frac{2,570}{50} = 51.40$

18) Answer: B.

Write the numbers in order: 10, 12, 18, 23, 24, 29, 37.

Since we have 7 numbers (7 is odd), then the median is the number in the middle, which is 23.

19) Answer: D.

Formula for the Surface area of a cylinder is:

$$SA = 2\pi r^2 + 2\pi r h \rightarrow 160\pi = 2\pi r^2 + 2\pi r(2) \rightarrow r^2 + 2r - 80 = 0$$

$$(r + 10)(r - 8) = 0 \rightarrow r = 8 \quad or \quad r = -10 \ (unacceptable)$$

20) Answer: A.

Let x be the number of years. Therefore, \$5,000 per year equals $5,000x$. Starting from \$36,000 annual salary means you should add that amount to $5,000x$.

Income more than that is: $I > 5,000x + 36,000$

Practice Tests 2

Section 1

1) Answer: A.

To find the sum of two whole numbers, line the numbers up and add digits from right.

$$\begin{array}{r} 53{,}126 \\ + \ 7{,}492 \\ \hline 60{,}618 \end{array}$$

2) Answer: D.

15% of \$420 is $0.15 \times 420 = 63$

3) Answer: C.

4 percent of $450 = \frac{4}{100} \times 450 = \frac{1}{25} \times 450 = \frac{1 \times 450}{25} = 18$

4) Answer: D.

The population is increased by 15% and 20%. 15% increase changes the population to 115% of original population. For the second increase, multiply the result by 120%.

$(1.15) \times (1.20) = 1.38 = 138\%$. 38 percent of the population is increased after two years.

5) Answer: A.

\$90 off is the same as 30 percent off. Thus, 30 percent of a number is 90.

Then: $30\% \ of \ x = 90 \rightarrow 0.3x = 90 \rightarrow x = \frac{90}{0.3} = 300$

6) Answer: C.

The result when 530 is divided by 7 is 75 with a remainder of 5. Multiplying $7 \times 75 = 525$ and $530 - 525 = 5$, which is the remainder.

7) Answer: C.

Four times of 15,000 is 60,000. One eighth of them cancelled their tickets.

One eighth of 60,000 equal 7,500 ($\frac{1}{8} \times 60{,}000 = 7{,}500$).

$52{,}500 (60{,}000 - 7{,}500 = 52{,}500)$ fans are attending this week.

8) Answer: D.

Change the numbers to decimal and then compare. $\frac{1}{5} = 0.2$;

0.83; $68\% = 0.68$; $\frac{5}{14} = 0.357$, Then: $\frac{1}{5} < \frac{5}{14} < 68\% < 0.83$

9) Answer: A.

First, find the number. Let x be the number. 160% of a number is 128, then: $1.6 \times x = 128 \Rightarrow x = 128 \div 1.6 = 80$; 45% of 80 is $0.45 \times 80 = 36$.

10) Answer: B.

$\frac{36}{32} = 1.125$. The only choice less than 1.125 is 1.065.

$1.065 < 1.125 < 1.152 < 1.250$

11) Answer: A.

1,800 out of 46,800 equals to $\frac{1,800}{46,800} = \frac{18}{468} = \frac{1}{26}$

12) Answer: D.

Dividing 192 by 16%, which is equivalent to 0.16, gives 1,200.

13) Answer: D.

The failing rate is 36 out of 45 $= \frac{36}{45}$. Change the fraction to percent: $\frac{36}{45} \times 100\% = 80\%$

80 percent of students failed. Therefore, 20 percent of students passed the exam.

14) Answer: B.

Use simple interest formula: $I = prt$ (I = interest, p = principal, r = rate, t = time)

$I = (22,000)(0.028)(7) = 4,312$

15) Answer: D.

To best compare the numbers, they should be put in the same format. The percent 0.453% can

be converted to a decimal by dividing 0.453 by 100, which gives 0.00453. $\frac{9}{20}$ can be converted

to a decimal by dividing 9 by 20, which gives 0.45. Now, all three numbers are in decimal

format. $0.450 > 0.0453 > 0.00453$ or $\frac{9}{20} > 0.0453 > 0.453\%$, which is choice D.

16) Answer: C.

The fraction $\frac{2}{5}$ can be written as $\frac{2 \times 20}{5 \times 20} = \frac{40}{100}$, which can be interpreted as forty hundredths, or

0.40.

17) Answer: D.

$\frac{3}{4} \times 32 = \frac{96}{4} = 24$

18) Answer: C.

The second digit to the right of the decimal point is in the

hundredths place and the third number to the right of the decimal point is in the thousandths

place. Since the number in the thousandths place of 0.2143, which is 4, is less than 5, the

number 0.2143 should be rounded down to 0.21.

19) Answer: D.

The amount of money that jack earns for one hour: $\frac{\$840}{35} = \$24$

Number of additional hours that he works to make enough money is: $\frac{\$2,100 - \$840}{1.5 \times \$24} = 35$

Number of total hours is: $24 + 35 = 59$

20) Answer: B.

$$2\frac{3}{4} - \frac{5}{8} = \frac{11}{4} - \frac{5}{8} = \frac{22-5}{8} = \frac{17}{8} = 2.125$$

Practice Tests 2

Section 2

1) Answer: D.

Simplify. $10x^4y^4(2x^2y^3)^3 = 10x^4y^4(8x^6y^9) = 80x^{10}y^{13}$

2) Answer: A.

The width of the rectangle is twice its length. Let x be the length. Then, $width = 5x$

Perimeter of the rectangle is 2 (width + length) = $2(5x + x) = 300 \Rightarrow 12x = 300 \Rightarrow x = 25$

Length of the rectangle is 25 meters.

3) Answer: B.

Plug in 59 for F and then solve for C.

$C = \frac{5}{9}(F - 32) \Rightarrow C = \frac{5}{9}(59 - 32) \Rightarrow C = \frac{5}{9}(27) = 15$

4) Answer: D.

Use Pythagorean Theorem: $a^2 + b^2 = c^2$

$10^2 + 24^2 = c^2 \Rightarrow 676 = c^2 \Rightarrow c = 26$

5) Answer: A.

$2x + 3y = 7$. Plug in the values of x and y from choices provided. Then:

A. $(2, 1)$ $2x + 3y = 7 \rightarrow 2(2) + 3(1) = 7 \rightarrow 4 + 3 = 7$ This is true!

B. $(5, 1)$ $2x + 3y = 7 \rightarrow 2(5) + 3(1) = 7 \rightarrow 10 + 3 = 13$ This is NOT true!

C. $(-2, 1)$ $2x + 3y = 7 \rightarrow 2(-2) + 3(1) = 7 \rightarrow -4 + 3 = -1$ This is NOT true!

D. $(3, -2)$ $2x + 3y = 7 \rightarrow 2(3) + 3(-2) = 7 \rightarrow 6 - 6 = 0$ This is NOT true!

6) Answer: B.

The diagonal of the square is 8. Let x be the side.

Use Pythagorean Theorem: $a^2 + b^2 = c^2$

$x^2 + x^2 = 8^2 \Rightarrow 2x^2 = 8^2 \Rightarrow 2x^2 = 64 \Rightarrow x^2 = 32 \Rightarrow x = \sqrt{32}$

The area of the square is: $\sqrt{32} \times \sqrt{32} = 32$

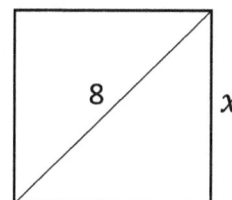

7) Answer: B.

To get a sum of 6 for two dice, we can get 5 different options:

$(1, 5), (5, 1), (2, 4), (4, 2), (3, 3)$

To get a sum of 4 for two dice, we can get 3 different options:

$(1, 3), (3, 1), (2, 2)$

Therefore, there are 8 options to get the sum of 4 or 6.

Since, we have $6 \times 6 = 36$ total options, the probability of getting a sum of 6 and 4 is 8 out of 36 or $\frac{8}{36} = \frac{2}{9}$.

8) Answer: C.

average (mean) $= \frac{\text{sum of terms}}{\text{number of terms}} \Rightarrow 55 = \frac{\text{sum of terms}}{40} \Rightarrow sum = 55 \times 40 = 2,200$

The difference of 77 and 37 is 40. Therefore, 40 should be subtracted from the sum.

$2,200 - 40 = 2,160$

mean $= \frac{\text{sum of terms}}{\text{number of terms}} \Rightarrow$ mean $= \frac{2,160}{40} = 54$

9) Answer: A.

In the stadium the ratio of home fans to visiting fans in a crowd is 4:9. Therefore, total number of fans must be divisible by 13: $4 + 9 = 13$.

Let's review the choices:

A. 11,219 $11,219 \div 13 = 863$

B. 13,453 $13,453 \div 13 = 1,034.85$

C. 24,729 $24,729 \div 13 = 1,902.23$

D. 56,840 $56,840 \div 13 = 4,372.31$

Only choice A when divided by 13 results a whole number.

10) Answer: B.

The perimeter of the trapezoid is 60 cm.

Therefore, the missing side (height) is $= 60 - 7 - 19 - 10 = 24$

Area of a trapezoid: $A = \frac{1}{2}h(b_1 + b_2) = \frac{1}{2}(24)(7 + 10) = 204$

11) Answer: D.

Use formula of rectangle prism volume.

$V = (\text{length})(\text{width})(\text{height}) \Rightarrow 1,800 = (25)(6)(\text{height})$

$\Rightarrow$ height $= 1,800 \div 150 = 12$

12) Answer: A.

Probability $= \frac{number\ of\ desired\ outcomes}{number\ of\ total\ outcomes} = \frac{14}{9+14+26+15} = \frac{14}{64} = \frac{7}{32}$

13) Answer: D.

average $= \frac{\text{sum of terms}}{\text{number of terms}} \Rightarrow$ (average of 9 numbers) $20 = \frac{\text{sum of numbers}}{9} \Rightarrow$ sum of 9 numbers

is: $20 \times 9 = 180$

(average of 6 numbers) $17 = \frac{\text{sum of numbers}}{6} \Rightarrow$ sum of 6 numbers is: $17 \times 6 = 102$

sum of 9 numbers – sum of 6 numbers = sum of 3 numbers $180 - 102 = 78 \Rightarrow$ average of 3

numbers $= \frac{78}{3} = 26$

14) Answer: B.

The probability of choosing the diamond and hearts is $\frac{26}{52} = \frac{1}{2}$

15) Answer: A.

Let x be the number of new shoes the team can purchase. Therefore, the team can purchase

$280x$.

The team had \$53,000 and spent \$45,000. Now the team can spend on new shoes \$8,000 at

most.

Now, write the inequality: $280x + 45,000 \leq 53,000$

16) Answer: D.

First, find the sum of four numbers.

average $= \frac{\text{sum of terms}}{\text{number of terms}} \Rightarrow 30 = \frac{\text{sum of 5 numbers}}{5} \Rightarrow$ sum of 5 numbers $= 30 \times 5 = 150$

The sum of 5 numbers is 150. If a sixth number that is greater than 60 is added to these

numbers, then the sum of 6 numbers must be greater than 210.

$150 + 60 = 210$. If the number was 60, then the average of the numbers is:

average $= \frac{\text{sum of terms}}{\text{number of terms}} = \frac{210}{6} = 35$

Since the number is bigger than 60. Then, the average of six numbers must be greater than

35. Choice D is greater than 35.

17) Answer: B.

Let L be the length of the rectangular and W be the width of the rectangular. Then,

$L = 8W - 3$

The perimeter of the rectangle is 102 meters. Therefore: $2L + 2W = 102$

$L + W = 51$

Replace the value of L from the first equation into the second equation and solve for W:

$(8W - 3) + W = 51 \rightarrow 9W - 3 = 51 \rightarrow 9W = 54 \rightarrow W = 6$

The width of the rectangle is 6 meters, and its length is:

$L = 8W - 3 = 8(6) - 3 = 45$

The area of the rectangle is: length × width = $45 \times 6 = 270$

18) Answer: C.

The ratio of boy to girls is 4:11. Therefore, there are 4 boys out of 15 students. To find the answer, first divide the total number of students by 15, then multiply the result by 4.

$105 \div 15 = 7 \Rightarrow 7 \times 4 = 28$

There are 28 boys and 77 (105 – 28) girls. So, 49 more boys should be enrolled to make the ratio 1:1

19) Answer: D.

Isolate and solve for x.

$\frac{4}{7}x + \frac{1}{6} = \frac{1}{2} \Rightarrow \frac{4}{7}x = \frac{1}{2} - \frac{1}{6} = \frac{1}{3} \Rightarrow \frac{4}{7}x = \frac{1}{3}$

Multiply both sides by the reciprocal of the coefficient of x. $(\frac{7}{4})\frac{4}{7}x = \frac{1}{3}(\frac{7}{4}) \Rightarrow x = \frac{7}{12}$

20) Answer: C.

Surface Area of a cylinder = $2\pi r (r + h)$,

The radius of the cylinder is 10 (20 ÷ 2) inches, and its height is 14 inches. Therefore,

Surface Area of a cylinder = $2\pi (10) (10 + 14) = 480 \pi$

"End"

www.ingramcontent.com/pod-product-compliance
Lightning Source LLC
Chambersburg PA
CBHW081327090426

42737CB00017B/3046

9 781636 200514